Palgrave Studies in Literature, Science and Medicine

Palgrave Studies in Literature, Science and Medicine is an exciting new series that focuses on one of the most vibrant and interdisciplinary areas in literary studies: the intersection of literature, science and medicine. Comprised of academic monographs, essay collections, and Palgrave Pivot books, the series will emphasize a historical approach to its subjects, in conjunction with a range of other theoretical approaches. The series will cover all aspects of this rich and varied field and is open to new and emerging topics as well as established ones.

More information about this series at
http://www.palgrave.com/gp/series/14613

Manon Mathias · Alison M. Moore
Editors

Gut Feeling and Digestive Health in Nineteenth-Century Literature, History and Culture

palgrave
macmillan

Editors
Manon Mathias
University of Glasgow
Glasgow, UK

Alison M. Moore
Western Sydney University
Penrith, NSW, Australia

Palgrave Studies in Literature, Science and Medicine
ISBN 978-3-030-01856-6 ISBN 978-3-030-01857-3 (eBook)
https://doi.org/10.1007/978-3-030-01857-3

Library of Congress Control Number: 2018957693

Cover credit: Wellcome Library no. 35068i

This Palgrave Macmillan imprint is published by the registered company Springer Nature Switzerland AG
The registered company address is: Gewerbestrasse 11, 6330 Cham, Switzerland

Acknowledgements

The editors gratefully acknowledge the following individuals and institutions for enabling this book to become a reality: Manon Mathias would like to thank the following organizations for supporting the symposium: the British Academy; the British Society for Literature and Science; the Society for the Study of French History; the Society for French Studies; and the University of Aberdeen School of Language, Literature, Music, and Visual Culture. Both editors thank the Palgrave anonymous reviewers for their helpful suggestions for improvement of this volume.

Alison M. Moore thanks Western Sydney University for providing a Women's Fellowship toward undergraduate teaching-relief in 2016 on the theme of 'Transcending the Science-Humanities Divide', and the British Academy for supporting her travel to the first 'Gut Feeling' symposium organized by Manon Mathias in Aberdeen, May 2017. Alison would also like to thank a termite's tiny intestinal micro-organisms that all danced an interactive wonder of inconceivable complexity before her eyes when she misread the laboratory instruction and put water on her slide instead of ethanol while studying physiology at the University of New England in 2013. This experience opened a whole new world of curiosity for her in microbiology and physiology, producing an openness to inter-disciplinary collaboration.

Contents

Contributors

Larry Duffy University of Kent, Canterbury, UK

Dorothy Johnson University of Iowa, Iowa City, IA, USA

Molly S. Laas University of Göttingen Medical School, Göttingen, Germany

Bertrand Marquer University of Strasbourg and Institut Universitaire de France, Strasbourg, France

Manon Mathias University of Glasgow, Glasgow, UK

Alison M. Moore Western Sydney University, Penrith, NSW, Australia

Catherine L. Newell University of Miami, Coral Gables, FL, USA

Tripp Rebrovick Harvard University, Cambridge, MA, USA

Emilie Taylor-Brown St Anne's College, University of Oxford, Oxford, UK

Cristiano Turbil Historian of Science and Medicine, King's College London, London, UK

Anne Vila University of Wisconsin–Madison, Madison, WI, USA

List of Figures

CHAPTER 1

The Gut Feelings of Medical Culture

Manon Mathias and Alison M. Moore

There is no doubt that the idea of gut health having systemic effects on human physiology is an ancient one, as countless health enthusiasts today pronounce with reference to supposed quotations from Hippocrates.[1] But in the nineteenth century, a critical mass of doctors, psychiatrists, novelists, artists, ethnographers, politicians, and religious leaders all began to consider that digestive function principally influenced both emotion and cognition, often dynamically engaging with such ideas across wide disciplinary divides. Such reflections on the relationship between the digestive system and the mind spread far beyond medical journals and clinical practices, representative of the new popular enthusiasm for scientific thought in nineteenth-century cultures, which resulted

[1] On the modern "invention" of Hippocratic medicine, see Cantor, "Introduction: The Uses and Meanings of Hippocrates"; King, "Hippocrates Didn't Write the Oath, so Why Is He the Father of Medicine?".

M. Mathias (✉)
University of Glasgow, Glasgow, UK

A. M. Moore
Western Sydney University, Penrith, NSW, Australia

M. Mathias and A. M. Moore (eds.), *Gut Feeling and Digestive Health in Nineteenth-Century Literature, History and Culture*, Palgrave Studies in Literature, Science and Medicine,
https://doi.org/10.1007/978-3-030-01857-3_1

in a broad range of genres through which evocation of the gut was imaginatively expressed.

This book is about that shifting focus on gut health in its relation to the mind and emotion, as epitomized in the English-language expression that refers to intuition as "gut feeling". But it is also a book about the emergence of this medical concept throughout literary, artistic, religious, nationalist, and ethnographic forms of representation. Embedded in this expression is the notion that the gut is the first part of the body to respond to the environment, producing an emotional sixth sense that then governs rational thought processes. This concept is not so foreign to our current physiological models in which the intestinal tract from mouth to anus is described as an exterior organ that mediates between the outside world and the internal viscera, with the intestinal epithelium providing an immunological gateway between self and other.[2] But our current medical models too have recently been undergoing an important shift toward even greater systemic emphasis on the importance of digestive function and intestinal microbiota.

In fact, we are living in an exciting moment for ideas about the digestive system since there is currently an explosion of scientific research in relation to the "gut-brain axis", suggesting that the gut microbiome may have a much larger role in human physiology than was previously realized, influencing not only gastrointestinal health itself, but also obesity, metabolic health,[3] the immune system,[4] cognitive function, mental illness, and neurodegenerative diseases.[5] In alternative health

[2] Silverthorn, *Human Physiology: An Integrated Approach*, 698.

[3] Belei et al., "The Relationship Between Non-Alcoholic Fatty Liver disease and Small Intestinal Bacterial Overgrowth Among Overweight and Obese Children and Adolescents"; Vrieze et al., "Transfer of Intestinal Microbiota from Lean Donors Increases Insulin Sensitivity in Individuals with Metabolic Syndrome"; and Müller et al., "Gastrointestinal Transit Time, Glucose Homeostasis and Metabolic Health: Modulation by Dietary Fibres."

[4] Tripathi et al., "Intermittent Hypoxia and Hypercapnia, a Hallmark of Obstructive Sleep Apnea, Alters the Gut Microbiome and Metabolome"; Armstrong and Armstrong, "Interactions of Gut Microbiota, Endotoxemia, Immune Function, and Diet in Exertional Heatstroke."

[5] Magnusson et al., "Relationship Between Diet-Related Changes in the Gut Microbiome and Cognitive Flexibility"; Nguyen et al., "Overview and Systematic Review of Studies of Microbiome in Schizophrenia and Bipolar Disorder"; and Westfalls et al., "Microbiome, Probiotics and Neurodegenerative Diseases: Deciphering the Gut Brain Axis."

communities over the past ten years, there has been an immense growth of popular interest in matters relating to the gut, as evidenced by the emergence of functional medicine in the USA, which has a strong focus on intestinal balance as the key to health and longevity. The opening pages of the work *Unconventional Medicine*, by the American functional health clinician Chris Kresser (one of the most influential figures of that movement) begins with the story of a child patient whose profound behavioral problems were found to correlate with severe intestinal bacterial imbalance, managed finally by a drastic change of diet.[6] The success of the bestselling book *Gut, the Inside Story* (2014) by the German gastroenterology PhD student, Giulia Enders, is a clear indication of how much questions of intestinal health are now seen as central among readers of books about health.[7] Popular interest in questions relating to the "gut-brain axis" proposed by increasing numbers of scientific researchers has been stimulated in recent years by the publication of articles on the web-based *Nature.com*[8] and in *The Lancet.*[9] It has now become common to hear even popular health enthusiasts cite the fact that there are more microbial cells in our bodies than the cells of the human body itself.[10] The implications of how we understand our "selves" is clearly suggesting a new vision of the human as multi-genomic, individualized as much by the composition of our gut micro-organisms as by our personalities, constituted as much by the life-forms living inside us as by our own genetic inheritance. There has also been a growth of interest specifically in ideas about the relationship between gut function and mental health, as a result of scientific work that has shown that the vagus nerve mediates neurotransmitter balance via the enteric nervous system, which was popularized in a series of recent articles in the web-based *Psychology Today.*[11]

[6] Kresser, *Unconventional Medicine.*

[7] Enders, *Gut.*

[8] https://www.nature.com/collections/dyhbndhpzv, accessed 12 October 2017.

[9] http://www.thelancet.com/journals/langas/article/PIIS2468-1253(17)30147-4/fulltext, accessed 12 October 2017.

[10] Yong, *I Contain Multitudes.*

[11] https://www.psychologytoday.com/blog/the-athletes-way/201708/the-microbiome-gut-brain-axis-relies-your-vagus-nerve, accessed 12 October 2017; https://www.psychologytoday.com/blog/brain-waves/201606/what-we-have-second-brain-the-microbiota-gut-brain-axis. http://www.thelancet.com/journals/langas/article/PIIS2468-1253(17)30147-4/fulltext, accessed 12 October 2017.

Central to this area of scholarship, which has boomed since the turn of the twenty-first century, is the importance of transnational team research, as biochemists, neurologists, endocrinologists, immunologists, gastroenterologists, and microbiologists from different parts of the world work together on questions such as the potential links between the gut and depression, cognitive decline, or mental illness. In our book, we have taken a similar approach toward our humanistic inquiry into nineteenth-century ideas about "gut feeling". Bringing together eleven scholars with wide interdisciplinary expertise, this volume examines literal and metaphorical digestion in a range of spheres in nineteenth-century life and asks why digestive health mattered for modern culture. Focusing on Northern America and Europe from the perspectives of literary studies; history of science and medicine; art history; politics; ethnography; and religious studies, we have targeted three central questions: Which meanings were given to the digestive system in the writings and culture of this period? How did these meanings differ between particular national contexts or transfer across them globally? How do these findings help us better understand the relationship between the body and the mind?

By looking at digestive health from a historical and cultural perspective, this volume offers distinctive insights into an emergent area of medicine that now so emphatically integrates the body, the mind, and the emotions. However novel the new gut-brain axis science may seem, in fact many individuals who engaged with medical thought in the nineteenth century drew intuitive conclusions of a similar nature. And just as today we see the gut taking center stage in popular health movements, so too in the nineteenth century it was an area of medicine particularly amenable to vulgarization. Of course, nineteenth-century explanations of the role of digestion in cognition, emotion, personality, and general health were speculative and often widely imaginative in contrast to the well-formulated mechanistic models suggested by scientists today.

But we should not fall prey to the easy tendency to mock the past as a means of elevating our present achievements. Without the creative cultural imagination of "gut feeling" elaborated in the culture of the long nineteenth century, it seems unlikely that medical researchers would ever have thought to look for ways in which digestion regulates other physiological functions, including those within the brain, or notice evidence of it doing so. Scientists do not operate in a vacuum but are embedded in cultural traditions and discourses, using both their left and right brain hemispheres to produce hypotheses that are triangulations of both deductive and inductive processing, both logical and intuitive reflection.[12]

Until now there has been little attempt to contextualize current ideas about the gut in scientific thought via a retrieval of older cultural discourses of "gut feeling". In fact, there has been strikingly little contribution to the emerging reimagination of gut health from scholars in the humanities. Since 2000, there has been a growing interest in food studies, a field populated mainly by historians and sociologists, which considers issues such as anorexia, obesity, and the social significance of eating practices.[13] Literary and cultural studies scholars have also shown great interest in corporality in recent decades, such as the consideration of literary texts themselves as "bodies", and the reading of the body as a textual surface or performative act.[14] Within medical history, there has been some interest in the history of the stomach, mainly focusing on gastric illness, constipation, and neurology.[15] But none of these areas has taken the operations of the digestive system as the central focus of inquiry. Over recent decades, interdisciplinary approaches have been successfully applied to the body in areas such as disability studies, history of sexuality, and the history of contagious disease,[16] but the time is now ripe to bring such a methodology to bear on digestive health.

Whereas many studies privilege either the medical or the cultural, this study draws attention to the dynamic flows between medical thought and culture, highlighting, in particular, the importance of emotion in questions of gut health, as reflected in the title's reference to "gut feeling". The volume suggests a new approach to corporality by looking at

[12] See McGilchrist, *The Master and His Emissary.*

[13] Forth and Carden-Coyne, eds., *Cultures of the Abdomen*; Wagner and Hassan, eds., *Consuming Culture in the Long Nineteenth Century*; Coveney, *Food, Morals and Meaning*; and Fischler, *L'homnivore.*

[14] Butler, *Bodies That Matter*; Grosz, *Volatile Bodies.* For examples of studies which explore the relations between text and body in particular authors, see, for example, Goellner and Murphy, eds., *Bodies of the Text*; Harrow, *Zola, The Body Modern*; and Georgopoulou, *The Body as Text in Shakespeare's Plays.*

[15] Miller, *A Modern History of the Stomach*; Whorton, *Inner Hygiene*; and Salisbury and Shail, eds., *Neurology and Modernity.*

[16] Roulstone, *Disability and Technology*; Fischer and Toulalan, eds., *Bodies, Sex and Desire from the Renaissance to the Present*; Bladerston and Guy, eds., *Sex and Sexuality in Latin America*; Moore, ed., *Sexing Political Culture in the History of France*; and Rütten and King, eds., *Contagionism and Contagious Diseases.*

the mid-lower body and its functions, but with a rigorous focus on the digestive system, rather than the important but much-studied questions of obesity or anorexia. In so doing, it advances the humanistic study of digestive health, an area of research which is today attracting major scientific attention, yet whose historical and cultural significance remains relatively unexplored.

By concentrating on this part of the body, our study challenges the existing literature on food in the nineteenth century, which has mostly examined gastronomy or social attitudes to eating rather than gastric health, regarded as puerile, unsavory, or simply irrelevant because it is assumed (incorrectly) to vary little across time and place. By opening up a cultural exploration of gastrointestinal health, the volume reveals its essential importance not only in current medicine and science but also in past politics, literature, art, ethnography, psychiatry, emotional health, intellectual processes, morality, and spirituality. As this list shows, the range of sources considered in the volume is broad, requiring an equally broad multi-disciplinary expertise, comprising literary studies, history of emotions, history of science and medicine, religious studies, politics, art history, and the history of anthropology and ethnography. Combined with this rich range of methodologies is a focus on changing views of the digestive system in nineteenth-century culture, as all chapters address the significance of gastric well-being in the period and its implications for our understanding of mind and body.

This is the first book then to consider digestive health in the nineteenth century from these mulitple combined angles. Each chapter focuses on a particular cultural context, time period, individual, text, visual material, or social theory, but all chapters address the meanings and significance of digestive health in modern culture. Given the centrality of the stomach in France in this period, especially in the fields of preventive medicine, alienist thought, gastronomy, and the novel—four fields in which the country was a leading force at this point—several essays take a French focus. These concerns, however, are part of a broader international discussion surrounding the place of the digesting body within modern society, and the volume examines the different permutations of this theme in other emerging industrial nations, in particular Northern America, Germanic Europe, Italy, colonial Australia, and Britain.

It was during the nineteenth century that the growth of sedentary forms of middle-class labor in these countries prompted new medical concerns about digestion and constipation, and research into gastrointestinal ailments focused on the links between the stomach and the brain both in the sense of cognitive capacity and mental health. Modern society's dualistic relationship with food and eating also began to solidify in our period of focus. The ethical and moral understanding of eating—derived from ancient Greek and Medieval Christian societies—continued to provide the basis for individual and political engagements with the notion of digestive health as a question of "hygiene". At the same time, biochemical, physiological, and nutritional discoveries rapidly changed the scientific understanding of digestion and began to rationalize the eating process with unprecedented levels of detail. This combination of both a rich moral and mechanistic understanding of food continues to influence our conflicted relationship with the digesting body today. It was also in the nineteenth century that cities throughout the industrializing world faced crises in their urban sewerage systems that were ill-equipped for the sudden population increases that were occurring. These crises produced "great stinks" that brought excrement into popular representation in new ways.

This was also the moment at which nation-states formed and expanded their power through imperial conquest, and medicine became increasingly intertwined with the state both at the metropolitan national level and in the colonies. Colonial power brought new and exotic food consumption possibilities to the burgeoning middle classes of the industrialized cities. Within these contexts, digestion became political, literally, and figuratively. Finally, the 1800s witnessed an explosion of literary, journalistic, and culinary writings on food, eating, and health, and a rapid rise in public literacy led to an expanding market for such materials. The volume focuses on these key trends by concentrating on digestion in relation to science, politics, and literature, providing a vital starting point for cultural and historical work on the emerging interdisciplinary field of gastrointestinal health. It will appeal to scholars and students of mind and body studies in literature, history, religious studies, the history of medicine and science, and visual studies, but also to health practitioners interested in the history of the digesting body.

Section One analyzes the importance of digestion on both a literal and figurative level within different spheres of political discourse, from French satirical prints to colonialist accounts of subjugated cultures. Tripp Rebrovick's essay offers an excellent example of the intersections between politics, literature, and science in the mid-nineteenth century and the importance of both the moral and the rational in questions of digestive health. Introducing readers to a newly discovered manuscript by poet Walt Whitman, Rebrovick explores Whitman's reshaping of humoral medicine in the language of nutrition science and argues that digestion was political for Whitman both in the sense of national American health and individual virtue. Bertrand Marquer's piece further examines the connections between digestion, politics, and literature by exploring the concept of the second brain and the importance of physiology in nineteenth-century French literature and ideology.

Alison Moore broadens the scope of the inquiry by considering the anal dimension of Freud's theories about civilization, arguing that these were heavily informed by nineteenth-century colonialist, ethnographical and anthropological studies in their discussion of excrement in "primitive" cultures. Her essay reveals the essential importance of reflections on defecation within Freudian thought but shows that this dimension has not hitherto been appropriately contextualized. Dorothy Johnson culminates this section on politics by demonstrating the role of digestive metaphors as a form of political critique in French art, influenced not only by caricatural prints from the French Revolution but also by British satirists.

The essays in Section Two offer new understandings of the connections between digestive processes and the process of writing, such as the metaphorical ingestion of medical discourses in naturalist novels or the links made between gastric ailments and the literary mind. The first essay in this section continues to draw inter-continental comparisons, as Emilie Taylor-Brown examines what was considered to be the peculiarly English phenomenon of dyspepsia, viewed as absent from France and Italy in this period due to their gastronomic culture. The essay engages explicitly with the question of emotion and sensibility through the notion of being "hangry", a recurring phenomenon in nineteenth-century debates on the relationship between the digestive and psychiatric systems. Anne Vila further probes the question of psycho-gastric links in an essay which explores the belief—held by physicians and literary writers in the late eighteenth and early nineteenth centuries—that the viscera might be at the very root of intellectual and aesthetic processes. Focusing on Voltaire, Diderot, and Balzac, Vila's essay shows the prevalence of mind-body connections especially in French conceptions of the writer at this point.

Manon Mathias develops the exploration of relations between literature, mind, and body by reading French novels from the second half of the century. Her essay compares these novels with contemporary health manuals and demonstrates their shared interest in the links between poor digestion and intellectual ability. Mathias argues that both categories of text seem to celebrate this link but shows that novelists Zola and Huysmans also drew on images of digestion and excretion to write about the release from physicality, in a paradoxical codependence. Many essays in the volume explore the slippage between literal and metaphorical digestion, and Larry Duffy furthers this exploration by looking at cases of actual indigestion in late nineteenth-century French novels and the concurrent, metaphorical ingestion of medical discourses by naturalist writers. His essay thus argues for the centrality of digestion in thinking about the development of discourses of knowledge.

Section Three, examining the place of digestion within nutrition science, comparative anatomy, and physiology, considers how the evolving understanding of digestion intertwined with the political, literary, and moral spheres. Opening this final section, Cristiano Turbil's essay on post-unification Italy considers the elaboration and dissemination of knowledge by revealing the importance of food science and digestion in Italian public health policy in the 1860s and 1870s. He uses this discussion as a basis for examining the communication of medical advice in the Italian popular context. Public health is similarly at the center of Molly Laas's essay, but this time in the context of 1830s New England and with a focus on the laborer. Laas examines the use of comparative anatomy by reformer Sylvester Graham and physician Luther Bell in debates surrounding "proper" diet for workers, debates which ultimately revolved around questions of social structure and national morality. It is therefore shown that digestive health was at the center not only of science and medicine but also of political and national discourse.

The final essay in the collection, Catherine Newell's discussion of Sylvester Graham's science, continues the focus on national health and also ties in with the other sections' focus on the moral and political dimension of digestion and the importance of writing and language in this debate. However, it also brings in the question of spirituality, as Newell considers Graham's diet and popular adherence to it as a form of spiritual belief. Her essay shows how Graham used physiological and nutritional science as a basis for theories of spiritual well-being and argues that this particular combination continues to influence and inform certain conceptions of health and wellness in the twenty-first century. Given the tight focus of the volume, which arises directly from a workshop aimed at publication, there are several interconnections between sections and within individual essays. For example, the political dimension is examined not only in Section One but also in many essays in Section Three, and in several essays, the scientific and the literary are naturally intertwined.

The volume has a broad geographic scope, with chapters focused on French culture (Marquer, Johnston, Vila, Mathias, and Duffy), several on the USA (Rebrovick, Laas, Newell), one on England (Taylor-Brown), one on Italy (Turbil), and one spanning Germanic, American, British, French, and Australian sources (Moore). The slight concentration of focus on French culture is a product of the topic in question: France was a striking example of fascination with questions of the gut during the nineteenth century, in part due to its strong materialist intellectual heritage, the importance of realist currents in its literary traditions, and the crises of its urban development and disease epidemics producing a sudden growth in medical knowledge of bacteria and intestinal illnesses. These crises were important drivers of the microbial discoveries of Louis Pasteur and the international medical shift toward a "germ theory" of disease in 1850–1870. The sudden changes in lifeways of the rapidly growing sedentary middle classes in this period also meant that those cultures with the most urban development faced new problems of intestinal health on a mass scale, hence the preponderance of representations from France, England, and the USA.

This collection of essays does not offer a history of only one part of the digestive system: Such studies already exist such as Ian Miller's *A Modern History of the Stomach* (2011), and James Whorton's *Inner Hygiene* (2000) with its focus on constipation; nor does it focus uniquely on the emergence of medical specialists of the gut such as W.F. Bynum's edited collection, *Gastroenterology in Britain* (1997).[17] Neither is this a book on food studies in the sense of cuisine, gastronomy, eating customs, eating disorders, or weight management. The focus is specifically on the digestive system: the volume concentrates on modern Western culture as an intellectual space in which a distinct set of practices and attitudes toward the gut began to solidify in the nineteenth century, particularly the tensions between the physical and the moral, emotional, or psychological. The book does not intend to offer a survey of

[17] Miller, *A Modern History of the Stomach*; Whorton, *Inner Hygiene*; and Bynum, ed., *Gastroenterology in Britain*.

attitudes toward digestive health in Western culture in general, and the approach is not one of comprehensive coverage. Rather, it highlights specific national contexts—especially France, Britain, and the USA—where strong preoccupations with questions of digestion appeared during the nineteenth century as well as considering the global exchanges across cultures in relation to such questions. Politics, science, and literature are shown to be important domains in which this preoccupation manifested itself, in tandem with each other and with different areas of nineteenth-century life including visual culture, spirituality, and emotion.

One of the aims of the book is to open up further lines of inquiry, and we gesture to the role of digestion in further areas such as sport, education, or the popular media that appear rich for exploration by other scholars. More work also needs to be undertaken on the history of gut-mind connections in non-Western cultures, particularly in Chinese, Vedic, and indigenous medical knowledges. A further fruitful line of scholarly inquiry would be to consider whether the rich literary engagements with gut health persists in today's cultural productions such as novels, artworks, and films. In short, we hope that this collection will inspire further scholars to engage in research on digestive well-being from a humanities perspective, paving the way for a genuinely interdisciplinary approach to gut health.

Bibliography

Armstrong, L. E., E. C. Lee, and E. M. Armstrong. "Interactions of Gut Microbiota, Endotoxemia, Immune Function, and Diet in Exertional Heatstroke." *Journal of Sports Medicine* (April 2018). https://doi.org/10.1155/2018/5724575.

Belei, O., L. Olariu, A. Dobresku, T. Marcovici, and O. Marginean. "The Relationship Between Non-Alcoholic Fatty Liver Disease and Small Intestinal Bacterial Overgrowth Among Overweight and Obese Children and Adolescents." *Journal of Pediatric and Endocrinology and Metabolism* 30, no. 11 (October 2017): 1161–1168. https://doi.org/10.1515/jpem-2017-0252.

Bladerston, Daniel, and Donna J. Guy, eds. *Sex and Sexuality in Latin America.* New York: New York University Press, 1997.

Butler, Judith. *Bodies That Matter: On the Discursive Limits of "Sex"*. London: Routledge, 1993.

Bynum, William F., ed. *Gastroenterology in Britain.* London: Wellcome Institute for the History of Medicine, 1997.

Cantor, David. "Introduction: The Uses and Meanings of Hippocrates." In *Reinventing Hippocrates*, edited by David Cantor, 1–16. London: Routledge, 2016.

Coveney, John. *Food, Morals and Meaning: The Pleasure and Anxiety of Eating*. London: Routledge, 2006.

Enders, Giulia. *Gut: The Inside Story of Our Body's Most Under-Rated Organ*. New York: Greystone Books, 2015.

Fischer, Kate, and Sarah Toulalan, eds. *Bodies, Sex and Desire from the Renaissance to the Present*. London: Palgrave, 2011.

Fischler, Claude. *L'homnivore. Le goût, la cuisine et le corps*. Paris: Odile Jacob, 2001.

Forth, Christopher E., and Ana Carden-Coyne, eds. *Cultures of the Abdomen: Dietetics, Digestion and Obesity in the Modern World*. New York: Palgrave Macmillan, 2005.

Georgopoulou, Xenia. *The Body as Text in Shakespeare's Plays: The Fashioning of the Sexes*. Lewiston, NY: Edwin Mellen Press, 2011.

Goellner, Ellen W., and Jacqueline Shea Murphy, eds. *Bodies of the Text: Dance as Theory, Literature as Dance*. New Brunswick, NJ: Rutgers University Press, 1995.

Grosz, Elizabeth. *Volatile Bodies: Toward a Corporeal Feminism*. Bloomington: Indiana University Press, 1994.

Harrow, Susan. *Zola, the Body Modern: Pressures and Prospects of Representation*. London: Legenda, 2010.

King, Helen. "Hippocrates Didn't Write the Oath, so Why Is He the Father of Medicine?" *The Conversation*, 3 October 2014. https://theconversation.com/hippocrates-didnt-write-the-oath-so-why-is-he-the-father-of-medicine-32334. Accessed 12 June 2018.

Kresser, Chris. *Unconventional Medicine: Join the Revolution to Reinvent Healthcare, Reverse Chronic Disease and Create a Practice You Love*. Austin, TX: Lioncrest Publishing, 2017.

Magnusson, K. R., L. Hauck, B. M. Jeffrey, V. Elias, A. Humphrey, R. Nath, A. Perrone, and L. E. Bermudez. "Relationship Between Diet-Related Changes in the Gut Microbiome and Cognitive Flexibility." *Neuroscience* 300 (August 2015): 128–140. https://doi.org/10.1016/j.neuroscience.2015.05.016.

McGilchrist, Iain. *The Master and His Emissary: The Divided Brain and the Making of the Western World*. New Haven, CT: Yale University Press, 2009.

Miller, Ian. *A Modern History of the Stomach: Gastric Illness, Medicine and British Society, 1800–1950*. London: Pickering and Chatto, 2011.

Moore, Alison M., ed. *Sexing Political Culture in the History of France*. New York: Cambria, 2012.

Müller, M., E. E. Canfora, and E. E. Blaak. "Gastrointestinal Transit Time, Glucose Homeostasis and Metabolic Health: Modulation by Dietary Fibres." *Nutrients* 10, no. 3 (February 2018): E.275. https://doi.org/10.3390/nu10030275.

Nguyen, T. T., T. Kosciolek, L. T. Eyler, R. Knight, D. V. Jeste. "Overview and Systematic Review of Studies of Microbiome in Schizophrenia and Bipolar Disorder." *Journal of Psychiatric Research* 99 (April 2018): 50–61. https://doi.org/10.1016/j.jpsychires.2018.01.013.

Roulstone, Alan. *Disability and Technology: An Interdisciplinary and International Approach*. London: Palgrave, 2016.

Rütten, Thomas, and Martina King, eds. *Contagionism and Contagious Diseases.* Berlin: Walter de Gruyter, 2013.

Salisbury, Laura, and Andrew Shail, eds. *Neurology and Modernity.* London: Palgrave Macmillan, 2010.

Silverthorn, Dee Unglaub. *Human Physiology: An Integrated Approach.* 6th ed. Boston: Pearson, 2013.

Tripathi, A., A. V. Melnik, J. Xue, O. Poulsen, M. J. Meehan, G. Humprey, L. Jiang, G. Ackermann, D. McDonald, D. Zhou, R. Knioght, P. C. Dorrestein, and G. G. Haddad. "Intermittent Hypoxia and Hypercapnia, a Hallmark of Obstructive Sleep Apnea, Alters the Gut Microbiome and Metabolome." *MSystems* 3, no. 3 (June 2018): E00020-18. https://doi.org/10.1128/msystems.00020-18.

Vrieze A., E. Van Nood, F. Holleman, J. Salojärvi, R. S. Kootte, J. F. Bartelsman, G. M. Dallinga-Thie, M. T. Ackermans, M. J. Serlie, R. Oozeer, M. Derrien, A. Druesne, J. E. Van Hylckama Vlieg, V. W. Bloks, A. K. Groen, H. G. Heilig, E. G. Zoetendal, E. S. Stroes, W. M. de Vos, J. B. Hoekstra, and M. Nieuwdorp. "Transfer of Intestinal Microbiota from Lean Donors Increases Insulin Sensitivity in Individuals with Metabolic Syndrome." *Gastroenterology* 143, no. 4 (October 2012): 913–916.e7. https://doi.org/10.1053/j.gastro.2012.06.031.

Wagner, Tamara S., and Tarin Hassan, eds. *Consuming Culture in the Long Nineteenth Century: Narratives of Consumption, 1700–1900.* Plymouth: Lexington Books, 2007.

Westfalls, S., N. Lomis, I. Kahouli, S. Y. Dia, S. P. Singh, and S. Prakash "Microbiome, Probiotics and Neurodegenerative Diseases: Deciphering the Gut Brain Axis." *Cellular and Molecular Life Sciences* 74, no. 20 (October 2017): 3769–3787. https://doi.org/10.1007/s00018-017-2550-9.

Whorton, James C. *Inner Hygiene: Constipation and the Pursuit of Health in Modern Society.* Oxford: Oxford University Press, 2000.

Yong, Ed. *I Contain Multitudes: The Microbes Within Us and a Grander View of Life.* London: Bodley Head, 2016.

Website Articles

https://www.nature.com/collections/dyhbndhpzv. Accessed 12 October 2017.

http://www.thelancet.com/journals/langas/article/PIIS2468-1253(17)30147-4/fulltext. Accessed 12 October 2017.

https://www.psychologytoday.com/blog/the-athletes-way/201708/the-microbiome-gut-brain-axis-relies-your-vagus-nerve. Accessed 12 October 2017.

https://www.psychologytoday.com/blog/brain-waves/201606/what-we-have-second-brain-the-microbiota-gut-brain-axis. Accessed 12 October 2017.

http://www.thelancet.com/journals/langas/article/PIIS2468-1253(17)30147-4/fulltext. Accessed 12 October 2017.

CHAPTER 2

The Great American Evil—Indigestion: Digestive Health and Democratic Politics in Walt Whitman

Tripp Rebrovick

In 1858, writing under the pseudonym Mose Velsor, Walt Whitman published a series of thirteen essays entitled "Manly Health and Training" that covered a range of subjects relating to nutrition, health, and regimen. This study situates Whitman's essays within dietetic debates of the mid-nineteenth century and focuses in particular on Whitman's response to the works of the German chemist Justus von Liebig and the American moral reformer Sylvester Graham. Liebig and Graham provided Whitman with the critical concepts that structured and informed his inquiry. Throughout the essays Whitman advanced two central and connected claims, first, that indigestion was "the great American evil" threatening the health of the American body politic and, second, that the healthiest diet consisted almost exclusively of meat. Whitman's concern about widespread indigestion tracked Liebig and Graham closely, but his resistance to vegetarianism, I contend, is best understood as a rejection

T. Rebrovick (✉)
Harvard University, Cambridge, MA, USA

M. Mathias and A. M. Moore (eds.), *Gut Feeling and Digestive Health in Nineteenth-Century Literature, History and Culture*, Palgrave Studies in Literature, Science and Medicine,
https://doi.org/10.1007/978-3-030-01857-3_2

of the Christian moralism, typified by Graham, that usually accompanied vegetarianism in the nineteenth century.

As I will explain shortly, mid-nineteenth-century Euro-American dietetic discourse was in the midst of a fundamental transformation. The classical tenets of humoral medicine were being slowly abandoned in favor of the principles articulated by the modern sciences of chemistry and biology. In light of this context, "Manly Health and Training" (hereafter MH&T) deserves a close study for a few reasons. First, it exemplifies and testifies to the fading power of humoral dietetics as an explanatory regime whose recommendations nevertheless continued to be authoritative even after its epistemic sources had lost their credibility. Whitman rehearses the standard humoral advice about moderate diet and exercise, but he justifies his recommendations by drawing on the new scientific understandings of the body. I focus on Whitman's liminal position throughout the essay in order to demonstrate and substantiate a point I have made elsewhere about the history of dietetic discourses.[1] A dietetic discourse, I argue following Foucault, constitutes its own unique object of knowledge and gives rise to distinct practices that assume the existence of that object.[2] Discourses, however, are not "hermetically sealed, monolithic truth regimes"[3] but structured sets of concepts and practices that can and do overlap. Sometimes they reinforce each other, sometimes they conflict. Whitman's essays exemplify this permeable characteristic of discursive formations. Second, MH&T clarifies many of the mysterious references to food and diet from Whitman's signed poetry and prose by revealing the poet's influences and interlocutors. With one prominent exception (to be discussed below), Whitman's references to food and diet in *Leaves of Grass* (1892) and *Democratic Vistas* (1871) are consistent with the ideas and prescriptions of MH&T. And third, MH&T explicitly connects proper diet, healthy digestion, and bodily fitness to broader debates concerning political virtues and formulae for successful democratic institutions. Whitman argued, in line with the dietary reformers of his day, that widespread indigestion was a pressing political problem preventing American citizens from reaching their fullest potential, perfection, and health. Digestion was political for Whitman

[1] Rebrovick, "The Politics of Diet: 'Eco-Dietetics,' Neoliberalism, and the History of Dietetic Discourses."

[2] Foucault, *The Archaeology of Knowledge; and, the Discourse on Language*.

[3] Taylor, "Foucault on Freedom and Truth," 279.

insofar as healthy bodies required proper diet, while democracy, in turn, required a host of healthy bodies able to fully digest their food. Whitman's essays demonstrate the political work of digestion and cultural understandings thereof.

Liebig and Graham

Before the emergence of nutrition science in the nineteenth century, the dominant dietetic discourse in European medicine was the system of humoral medicine.[4] According to the tenets of humoral medicine, all entities in the cosmos—food and the human body included—were composed of four specific properties or qualities (hot, cold, wet, and dry). Humoralism thereby understood digestion as the assimilation of the qualities of the food by the eater.[5] With the rise of nutritionism, however, food and the human body came to be understood as sets of components or nutrients, i.e., collections of chemicals rather than collections of qualities.[6] Chemical constituents (eventually encompassing carbohydrates, proteins, fats, vitamins, and minerals) replaced the four humors (blood, yellow bile, black bile, and phlegm) as the building blocks of both food and the body. Human bodies ceased to be balanced organic systems assimilating the qualities of their foods and became instead motors or machines requiring food as fuel.

A central figure in effecting this transition was the German chemist Justus von Liebig. In particular, Liebig revolutionized the fields of plant and animal nutrition, first, by arguing that plants required minerals and not just organic matter to grow, and, second, by describing ideal human diets in terms of their nutritional or chemical components rather than in terms of their agreement with the temperament or constitution of the eater. In his text *Chemistry in Its Application to Agriculture and Physiology* (1842), Liebig argued that plants required minerals, especially nitrogen, in order to grow.[7] In the *Familiar Letters on Chemistry* (1843), Liebig's more widely circulated and popular text, he explained his new

[4]Temkin, *Galenism: Rise and Decline of a Medical Philosophy.*

[5]Shapin, "'You Are What You Eat.'"

[6]Scrinis, "On the Ideology of Nutritionism"; Scrinis, *Nutritionism*; and Shapin, "'You Are What You Eat.'"

[7]Liebig, *Chemistry in Its Application to Agriculture and Physiology.*

theories of digestion to general audiences.[8] Liebig's conclusions about plant nutrition paved the way for industrial agriculture and its use of chemical fertilizers, while his theories of animal nutrition helped end the reign of humoral dietetics and cement nutritionism as the dominant dietetic regime. In order to contextualize Liebig's innovations, I will now survey in more detail the dominant theory of digestion that prevailed in humoral medicine.

According to the humoral doctors, Galen especially, digestion is a linear sequence that includes the heating of food within the stomach in a process typically known as concoction. In this model, indigestion and other stomach problems are explained by a dearth of natural heat in the stomach, either because the body's heat is being used elsewhere for another activity like thinking or walking, or because the body's overall stores of heat are low due to old age or disease. Even Kant accepted as fact that "intellectual work diverts vital energy from the stomach and bothers it."[9] According to the humoral model, the body is endowed with a finite amount of natural heat and vital spirits to consume over its lifetime. Moderating food intake, then, could prolong life by limiting the amount of heat used to consume and digest food.

With Liebig, however, digestion ceased to be a linear sequence. Instead, digestion became bifurcated into two distinct, parallel processes, one supplying nutrients for the reproduction of bodily tissue, and the other supplying energy by means of respiration. In *Familiar Letters*, Liebig divided "all the substances which constitute" food "into two great classes."[10] The first class contributes to the "nutrition and reproduction of the animal body,"[11] while the second helps create "animal heat" by combusting with the oxygen in the air.[12] In Liebig's system, rather than consuming or using up the body's vital spirits, digestion created new heat for the body to use. The body's store of heat was no longer finite but was replenished at each meal. Rather than analogizing the stomach to a candle, as the humoral authors often did, Liebig

[8] Liebig, *Familiar Letters on Chemistry and Its Relation to Commerce, Physiology, and Agriculture.*

[9] Kant, *The Conflict of the Faculties* (*Der Streit Der Fakultäten*), 199.

[10] Liebig, *Chemistry in Its Application to Agriculture and Physiology*, 64.

[11] Ibid.

[12] Ibid., 70.

compared the stomach to "a furnace which we supply with fuel."[13] Liebig's furnace metaphor is especially significant for its comparison of the human body to an artificial machine rather than an organic body. Understood as a machine, the human body's powers and capabilities could be known and studied through quantification and thereby be manipulated in order to achieve maximum productivity. According to the mechanical model of the body, fatigue resulted from a lack of energy in the body rather than a constitutional or moral defect. Liebig's nutritional chemistry, in this manner, helped complete the modern rejection of Aristotelian teleology and natural philosophy.

Of equal significance, moreover, was Liebig's focus on the components or nutrients of food, rather than the qualities, and the consequent emphasis on the quantification of proper diets.[14] Liebig's divisions of foods into classes eventually morphed into the contemporary division between proteins, carbohydrates, and fats (vitamins and minerals were not discovered until the 1920s), and quantification of these components was quickly incorporated into diet recommendations. By the turn of the twentieth century, definite amounts of specific nutrients were recommended in popularly distributed nutrition tables.[15] This sort of dietary knowledge vastly expanded the possibilities of what Foucault has called "biopolitics."[16] With the discovery of various components of foods, populations could be known, represented, and mapped according to their quantifiable nutritional needs. A population, for example, can be known and represented as an aggregate number of calories needed per day. Lastly, the reduction of food to chemical constituents enabled dietary advice to be targeted to human bodies in general in contrast to the humoral emphasis on tailoring diet to individual temperament or constitution.

Liebig, moreover, was known not only for his popular texts, but also for the commodities that bore his name. After experimenting with different kinds of plant fertilizer, Liebig turned his attentions to commercial food products. In the 1860s, his company began to sell and market one

[13] Liebig, *Familiar Letters on Chemistry and Its Relation to Commerce, Physiology, and Agriculture*, 74.

[14] Rebrovick, "The Politics of Diet: 'Eco-Dietetics,' Neoliberalism, and the History of Dietetic Discourses"; Scrinis, *Nutritionism*; and Shapin, "'You Are What You Eat.'"

[15] Cullather, "The Foreign Policy of the Calorie"; Levine, *School Lunch Politics*.

[16] Foucault, *Security, Territory, Population*.

of the late-nineteenth century's most well-known commodities: Liebig's *Extract of Meat.*[17] By the 1840s and 1850s, doctors in Britain and the USA were prescribing "beef tea," sometimes known as *extractum carnis*, for all sorts of illnesses and maladies.[18] These beef teas were prescribed to convalescent patients in hospitals, in prisons, and elsewhere. According to humoral medicine, drinking beef tea enabled the patient to absorb and assimilate the qualities conventionally ascribed to beef: strength, vitality, and warmth. Beef teas were said to contain the essential qualities of beef in a liquid form. Nutritionism, however, proffered beef *extract* rather than beef *tea*. Rather than steeping beef in hot water in order to make a tea or a stock, Liebig's company reduced beef products down until they formed a condensed paste.[19] Whereas beef tea contained the properties of beef in an easily digestible form, beef extract contained the vital nutrients of meat separated from the supposedly unnecessary parts. Liebig's Extract was originally marketed as a nutritional supplement, although within twenty years those health claims had been so thoroughly debunked that Liebig's company was forced to adopt a new marketing strategy selling the product solely as a flavoring agent.[20]

Before proceeding to the text of MH&T, however, it is necessary to introduce one more important figure from American dietetics in the nineteenth century: Sylvester Graham. Graham's name is familiar to twenty-first-century eaters because of the crackers that bear his name, but Graham was first and foremost a moral reformer. Graham lectured around the USA on various topics, but he was especially known for his treatise on bread, his vegetarianism, and his anti-onanism. Graham

[17]Brock, *Justus von Liebig*; Finlay, "Quackery and Cookery"; and Finlay, "Early Marketing and the Theory of Nutrition: The Science and Culture of Liebig's Extract of Meat."

[18]Finlay, "Quackery and Cookery"; Finlay, "Early Marketing and the Theory of Nutrition: The Science and Culture of Liebig's Extract of Meat"; and Rosenberg, "And Heal the Sick."

[19]"Workers pressed animal flesh into a pulp with steam powered iron rollers. This pulp was then dropped into hot water, and steamed for one hour. The liquid was then allowed to ooze off into another vat, where workers removed the fat, and finally used heat and a vacuum evaporation pan to reduce the juice into a thick gravy. After several hours of cooling, the liquid was refiltered, packaged into small tins or pots, and reading for export." Finlay, "Early Marketing and the Theory of Nutrition: The Science and Culture of Liebig's Extract of Meat," 58.

[20]Finlay, "Quackery and Cookery."

advocated both a strictly vegetarian diet and an ascetic lifestyle, and he blamed bad diet for both disease and immorality.[21] In particular, Graham thought that eating meat (along with other problematic foods such as coffee, tea, spices, and pastries) inflamed the passions and promoted the vice of masturbation. Graham also faulted meat-eating for the prevalence of indigestion (or dyspepsia) in mid-nineteenth-century America.[22] Whitman was familiar with Graham and mentions "Grahamites" in letters from the 1840s.[23] Many of Whitman's dietary proscriptions resemble Graham's, but as I will show, he departed from Graham in certain crucial respects by drawing on Liebig instead.

"Manly Health and Training"

Whitman's signed material, both his prose and his poetry, contains only a few alimentary metaphors, but Zachary Turpin has recently discovered a nearly fifty-thousand-word series of articles by Whitman about dietetics. Turpin describes the essays as "Whitman's forgotten, book-length guide to living healthily in America."[24] For Turpin, the series is "part guest editorial, part self-help column" and could be described as anything ranging from "an essay on male beauty, a chauvinistic screed, a sports memoir, a eugenics manifesto, a description of New York daily life, an anecdotal history of longevity, or a pseudoscientific tract."[25] Although the essays do indeed straddle numerous genres, they also resemble conventional dietetic treatises from the mid-nineteenth century and respond to prominent dietetic reformers in the antebellum USA, especially Liebig and Graham. By situating Whitman's essays within dietetic debates of the mid-nineteenth century, I aim to clarify or explain propositions that otherwise seem random, curious, or out of place.

[21] Graham, *A Lecture to Young Men on Chastity*.

[22] Graham summarizes his conclusions as follows: "Improper diet—the free use of flesh, with more or less of stimulating seasonings and condiments, together with coffee, tea, rich pastry, and compounded and concentrated forms of food; and too often, chewing and smoking tobacco, and drinking wine and other intoxicating liquors;—all of which unduly stimulate and irritate the nervous system, heat the blood, and early develop a preternatural sensibility and prurience of the genital organs" ibid., 152–153.

[23] Whitman, *The Correspondence*, vol. 7, 5.

[24] Turpin, "Introduction to Walt Whitman's 'Manly Health and Training,'" 148.

[25] Ibid., 149.

Turpin documents the texts and medical journals from which Whitman drew extensively, highlighting the water cure movement and the phrenologists,[26] but Whitman was also quite familiar with Liebig. He had reviewed Liebig's *Chemistry in its Application to Agriculture and Physiology* in the *Brooklyn Daily Eagle* and written the following of the book's achievements: "Chemistry! The elevating, beautiful study! … Chemistry—that involves the essences of creation, and the changes, and the growths, and formations and decays, of so large a constituent part of the Earth."[27] As I noted in the previous section, Liebig had argued that minerals along with dead organic matter became a part of the soil and were reabsorbed by living organisms, especially plants. By eating these plants, humans and animals absorbed the same matter. All life was therefore part of a large, ever-repeating recycling of atoms. Whitman would eventually incorporate the language of this review directly into one of his poems, "This Compost," with the line "What chemistry!"[28]

Near the beginning of MH&T, Whitman equated manly health with "a perfect body, perfect blood—no morbid humors, no weakness, no impotency or deficiency or bad stuff in him."[29] A healthy body is "all running over with animation and ardor, all marked by herculean strength, suppleness, a clear complexion, and the rich results (which follow such causes) of a laughing voice, a merry song morn and night, a sparkling eye, and an ever-happy soul!"[30] He then proceeded to outline a program of health that covered diet, exercise, daily habits, and even clothes, but for Whitman a healthy body required, above all else, a healthy stomach and proper digestion. He stressed that "regular and complete assimilation, digestion, and excretion are the primal requisites of health."[31] Proper digestion was so important that "all other rules and requisites may be attended to, but if the stomach be out of order … all will be of no avail."[32] Diseases of the gut, moreover, threatened not just

[26] Ibid., 159.

[27] Whitman, "Review of Liebig's Chemistry in Its Application to Agriculture and Physiology."

[28] Whitman, *Poetry and Prose*, 496.

[29] Whitman, "Manly Health and Training, With Off-Hand Hints Toward Their Conditions," 184.

[30] Ibid.

[31] Ibid., 209.

[32] Ibid., 210.

individual bodies, but by extension, the American body politic as well. Whitman thus announced his opponent as "THE GREAT AMERICAN EVIL—INDIGESTION."[33] Sometimes he referred to the condition as dyspepsia, but he shared with Graham and their contemporaries a concern that American stomachs had become overtaxed, overstuffed, and exhausted. Whitman even suggested that "the general illness, perhaps transcending all the rest, is dyspepsia."[34] Whitman's solution to the great evil of indigestion and the enfeebled bodies it produced, however, recast conventional humoral recommendations in a nutritionist cosmos.

A practitioner or student of humoral medicine would find familiar material in MH&T, for it incorporated several tenets of that medical tradition: the vocabulary of the four humors; a focus on the "six non-naturals"; an emphasis on regimen and moderation; and the need for tailoring diet to individual circumstances. What can seem like a jumbled stream of consciousness to twenty-first-century readers actually had its own coherence, mirroring a conventional sequence of recommendations and topics that had remained relatively stable for nearly two millennia. But the story does not end there, for the essays that compose MH&T also drew from the burgeoning nutritionist discourse by referring, in turn, to the body as a machine, to the interchangeability of blood and tissue, and to the energeticist understanding of fatigue. I will now discuss each of these points in more detail before turning to Whitman's espousal of an all-meat diet and concluding with a brief discussion of his signed poetry and prose.

Although he did not describe the body as the balance of four humors, Whitman did diagnose certain bodily conditions as "choleric"[35] or "melancholic"[36] and associated health with the absence of "morbid humors."[37] Whitman did not rely on the cosmology of the humoral system, but he used its vocabulary and largely adopted its recommendations. For instance, the sequence of topics covered over the course of the essays closely resembles what were traditionally known as "the six non-naturals." The non-naturals were the six items not wholly subject to nature but which could be manipulated or controlled to some extent

[33] Ibid.

[34] Ibid., 209.

[35] Ibid., 185.

[36] Ibid., 271.

[37] Ibid., 184.

by human effort. They were: air, food and drink, motion and rest, sleep and watch, evacuations and retentions, and the passions of the mind.[38] Most cures recommended by a humoral physician would involve manipulating one or more of the six non-naturals.[39] Whitman referenced the six by noting that the training he advocated "does not consist in mere exercise. Equally important with that are the diet, drink, habits, sleep, &c,. Bathing, [and] the breathing of good air."[40] Whitman substituted bathing for evacuations and retentions and devoted a significant portion of the essays to bathing rituals, but otherwise adopted the commonly rehearsed list. In a different article of the series, Whitman kept four items from the list but substituted friends and employment for air and (again) for evacuations; he recommended "early rising, early to bed, exercise, plain food, thorough and preserving continuance in gently-commenced training, the cultivation with resolute will of a cheerful temper, the society of friends, and a certain number of hours spent everyday in regular employment."[41]

Throughout the sequence, Whitman recommended a strict regimen, although he preferred the word training. He emphasized that training required a daily routine covering the whole of life: "Training! In its full sense, it involves th*e entire science of manly excellence, education, beauty, and vigor*—nor is it without intimate bearings upon the moral and intellectual nature."[42] Humoral doctors described a daily regimen as the key to health, and its cardinal feature was moderation.[43] Throughout MH&T, Whitman endorsed "the simplest and most natural diet"[44] along with "simple and hearty food, and no condiments."[45] "Any article craved

[38] Coleman, "Health and Hygiene in the Encyclopédie"; Rather, "The 'Six Things Non-Natural': A Note on the Origins and Fate of a Doctrine and a Phrase."

[39] Shapin, "How to Eat Like a Gentleman: Dietetics and Ethics in Early Modern England"; Siraisi, *Medieval & Early Renaissance Medicine.*

[40] Whitman, "Manly Health and Training, With Off-Hand Hints Toward Their Conditions," 196.

[41] Ibid., 189.

[42] Ibid., 195. Emphasis original.

[43] Foucault, *The History of Sexuality. Vol. 2: The Uses of Pleasure*; Shapin, "How to Eat Like a Gentleman: Dietetics and Ethics in Early Modern England"; and Shapin, "'You Are What You Eat.'"

[44] Whitman, "Manly Health and Training, With Off-Hand Hints Toward Their Conditions," 197.

[45] Ibid., 212.

by the appetite, and not of essential importance to be prohibited," he counseled, "may be allowed in moderation."[46] To be avoided instead was "indulgence in any great variety of dishes at the same meal."[47] To ensure that one's diet remained moderate, Whitman suggested "eat[ing] according to your appetite, of one dish."[48] The abandonment of this principle caused disease or other symptoms of ill-health. "The effect of these highly artificial dishes," Whitman wrote, "is to stimulate and goad the appetite, distend the stomach, thin the blood, and prepare the way for some form or other of disease."[49] Moderation, health, and virtue were co-constitutive.

Lastly, Whitman adopted the humoral practice of qualifying every general rule or principle with the need to tailor it to individual constitution, temperament, or circumstance.[50] He noted, for example, that "we are willing to admit that our exact statements with regard to diet ... are to be often modified to suit cases, tastes, &c, each one for itself."[51] Whitman summarized his advice with recommendations that had been rehearsed for more than two millennia: "In general terms, avoid what disagrees with you; for there are, to every individual case, certain rules which apply to it alone."[52] The use of the word disagreement is telling, for agreement was the metric according to which humoral medicine judged the healthiness of food for a particular individual. By the turn of the end of the nineteenth century, however, quantitative precision replaced agreement as the determinant of health.

Although humoral metaphors were more common than nutritionist metaphors over the course of the thirteen essays, Whitman deployed nutritionist imagery in two significant places. He described the body as "a machine calculated to produce force"[53] and he diagnosed an

[46] Ibid., 307.

[47] Ibid., 197.

[48] Ibid., 201.

[49] Ibid., 269.

[50] Foucault, *The History of Sexuality. Vol. 2: The Uses of Pleasure*; Shapin, "How to Eat Like a Gentleman: Dietetics and Ethics in Early Modern England"; and Shapin, "'You Are What You Eat.'"

[51] Whitman, "Manly Health and Training, With Off-Hand Hints Toward Their Conditions," 203.

[52] Ibid., 306–307.

[53] Ibid., 225.

individual lacking certain feelings as lacking "electricity" in the body.[54] Anson Rabinbach has shown that in the nineteenth century, as a result of the new sciences of thermodynamics, energeticist physics, and organic chemistry, the human body began to be compared to a motor or a machine.[55] Whitman's claim that the "electric" body can "produce force" in a "calculated" manner drew upon the new sciences of energy.

The essays also adopted Liebig's understandings of the interchangeability between blood and bodily tissue. Liebig argued that the components of foods were the same as the components of blood, and further, that the constituents of blood were the same as the constituents of bodily tissues. Those tissues absorbed nutrients from the blood as needed for restoration. Whitman tracked Liebig closely when he described "*the deep, interior, sane cure of the whole quality of the blood and the tissue it forms*, which make the body."[56] Liebig posited the commensurability of blood, tissue, and the whole body by reducing all entities to their chemical constituents. Whitman's rejection of particular medicines, moreover, followed from this understanding of the workings of blood: "Medicines … may give temporary relief—but the only effectual medicine lies in *the entire purification and renovation of the life of the body, the blood.*"[57] The humoral system also gave blood a prominent place as one of the four humors, but the "life of the body" was associated with "natural heat," the "vital spirits," or "radical moisture" rather than blood itself. The prominence of blood in Whitman's essay and its commensurability with tissue connected Whitman's dietary advice to nutritionist understandings of the body.

The essays in MH&T, furthermore, interpreted indigestion as the fatigue or overworking of the stomach. Before the emergence of energeticist physics and nutrition science, fatigue or tiredness was usually explained and condemned in moral terms.[58] Those who were too tired to work were lazy or immoral. After the concept of the energy balance became widespread, however, fatigue was re-explained as the use of

[54] Ibid., 221.

[55] Rabinbach, *The Human Motor*.

[56] Whitman, "Manly Health and Training, With Off-Hand Hints Toward Their Conditions," 276.

[57] Ibid., 289.

[58] Rabinbach, *The Human Motor*.

more energy in a day than the body was capable of creating for itself.[59] Fatigue was caused by overwork or undernutrition and not by constitutional lassitude. Whitman adopted this new understanding of fatigue when he argued that health required one to keep the digestive "system from wearing out and stagnating upon itself."[60] In a section focusing on bodybuilders and prizefighters, Whitman castigated athletes who tried to change their bodies too quickly, noting that "the injurious fatigue of crowding so much into so small a space, destroys and reacts upon itself."[61] The most telling application of the new theory of fatigue, however, can be seen in Whitman's promulgation of an exclusively meat-based diet.

Throughout the essays, Whitman recommended "*an almost exclusive meat diet.*"[62] Here he was quite blunt: "*Let the main part of the diet be meat, to the exclusion of all else.*"[63] Although Whitman's defense of meat-eating is not unusual in the history of dietetics, his defense of eating only or primarily meat certainly is. Whitman's resistance to vegetarianism, I contend, is best understood as a rejection of the Christian moralism that inevitably accompanied vegetarianism in the nineteenth century, especially in the figure of Sylvester Graham. In order to make his case for meat-eating, however, Whitman endorsed a humoral practice used for treating convalescent patients but justified it with a new rationale in light of Liebig's theories.

Whitman recapitulated most of Graham's recommendations, but he offered directly opposing advice on the issue of vegetarianism. Three arguments from Graham were adopted nearly verbatim by Whitman. First, Graham rejected the use of spices and condiments in food and prohibited drinking hot beverages like coffee and tea.[64] Second, Graham advocated that young men "should rise early—bathe the head and neck freely, and the whole body once a day—use friction over the whole surface—and exercise actively and vigorously."[65] And third, Graham worried

[59] Ibid.

[60] Whitman, "Manly Health and Training, With Off-Hand Hints Toward Their Conditions," 243.

[61] Ibid., 281.

[62] Ibid., 268. Emphasis original.

[63] Ibid., 269. Emphasis original.

[64] Graham, *A Lecture to Young Men on Chastity*, 152–153.

[65] Ibid., 156.

that "over-exercising the brain" could damage the stomach.[66] Whitman wholly adopted these three tenets of Graham's, including the use of the "flesh-brush"[67] and the warning against "too much brain action and fretting."[68] But when Whitman turned to his own version of a simple diet, he explicitly distanced himself from Graham's "plain, simple, un-stimulating, vegetable and water diet."[69] Whitman contrasted his diet implicitly but clearly with Graham's: "the simplest and most natural diet is the best, and lest we be misunderstood, we specify that we do not mean a vegetarian or water-gruel diet, but one of strengthening materials, beef, lamb, &c, and that fruits, wines, and the like are not to be excluded."[70] Whitman's rejection of the vegetable and water diet was swift and succinct: "we have seen New England and New York vegetarians, gaunt, hard, melancholy, and unhappy looking persons, that looked like anything else than a recommendation of their doctrine—for that is the proof, after all."[71] He acknowledged that some individuals could be healthy on a vegetarian diet, but argued against its widespread adoption in the USA.

What drove Whitman to this curious recommendation? I contend that two factors can best explain it: first, Whitman's rejection of Christian asceticism and his consequent embrace of sexuality, and second, Whitman's concern, drawn from Liebig's account of digestion, that a too-varied diet was overtaxing the American stomach.

The poet of *Leaves of Grass* not only embraced the body and its pleasures but also rejected religions and moral reformers. Erotic and sexual imagery saturated the poems so much so that the first editions were viciously attacked and often banned outright. In MH&T, Whitman took a similar position, praising the male physique's sexual magnetism and explicitly rejecting Graham's tendency to be "ashamed of the carnal

[66] Ibid., 163.

[67] Whitman, "Manly Health and Training, With Off-Hand Hints Toward Their Conditions," 199.

[68] Ibid., 296.

[69] Graham, *A Lecture to Young Men on Chastity*, 156.

[70] Whitman, "Manly Health and Training, With Off-Hand Hints Toward Their Conditions," 197.

[71] Whitman, *Walt Whitman's Songs of Male Intimacy and Love*, 234.

body."[72] Although Whitman, like Graham, endorsed an anti-onanist position, he did so using a humoral rather than a Christian rationale. Rejecting Graham's asceticism, Whitman praised "the curious magnetism" that accompanied "manly physique."[73] In maintaining this magnetic presence, however, "a main part doubtless lies in the department of sexuality.... In the department we speak of, a reserved stock of vital energy, we say, marks the *man*; and he who gives himself up, by its undue exhaustion ... must bear the miserable consequences."[74] In other words, like digestion, sexual activity used up the body's finite amount of vital energy—its "reserved stock." The body's sexual economy in MH&T still operated according to humoral principles; a limited amount of spirits or heat were able to be consumed over the course of life. Too much sexual activity drained the vital spirits in the same manner that too much digestion did. Whitman, then, aimed to preserve and strengthen sexual energies through moderation rather than dampen and eliminate them, as his contemporaries most often did. Whitman recommended a regimen designed to cultivate the "curious magnetism" that healthy and robust (male) bodies exhibited. Aware of this distance from his peers, Whitman noted that "this subject is probably new to you, reader, treated in this way."[75] Whitman's impulse to reject vegetarianism stemmed from his denunciation of Christian moralism.

Whitman's preference for meat is further explained by his desire that individuals consume just one type of food at every meal. With "a simple meat diet,"[76] Whitman maintained, "the digestive organs would have more than half the labor (agonizing labor, it often is) withdrawn from them, and the blood relieved from an equally great amount of noxious deposit, which, under the present system, is thrown into it."[77] Meat for Whitman was the one item that, in itself, could provide sufficient nutriment. Given the predominance of beef teas in the 1840s and 1850s, it is not surprising that Whitman turned to beef. But the analysis of the stomach's fatigue relied on Liebig's understanding of digestion.

[72] Whitman, "Manly Health and Training, With Off-Hand Hints Toward Their Conditions," 269.

[73] Ibid., 223.

[74] Ibid., 225.

[75] Ibid.

[76] Ibid., 268.

[77] Ibid., 269–270.

Subsisting solely on meat relieves the stomach of "agonizing labor" and the presence of "noxious" substances that could neither be assimilated nor combusted in order to make energy. Those who ate too many foods would "distend the stomach, and bloat themselves with quantities of trash, to worry the digestion, thin the blood, and return, sooner or latter, in lassitude, headache, constipation, or a fever, or some other attack of illness."[78] The stomach of the overeater "has been overloaded with all sorts of indigestible solids and injurious liquids … digestion is seriously impaired."[79] "Excessive stimulus"[80] of the stomach produced fatigue and forced the body to consume more energy on a daily basis than it could produce by combusting food, and "the stomach needs rest as much as the other parts of the system."[81] An all-meat diet ensured health, not because of agreement, but because of efficiency.

Whitman's embrace of carnivorous principles followed from both his rejection of ascetic vegetarianism and his conclusion that American stomachs experienced indigestion because of diets that were too varied, too artificial, and too large. A diet of vegetables alone was emaciating; a diet of meat and vegetables was overtaxing; a diet of primarily meat was just right. This curious position of Whitman's stems from the fundamental shift dietetic culture was undergoing in the mid-nineteenth century. His text straddled the line between two dietetic discourses and provides a compelling example of an ongoing discursive transformation.

Alimentary Metaphors in *Leaves of Grass* and *Democratic Vistas*

Whitman's preference for beef, however, did not stay confined to the pseudonymous essays, but worked its way into *Leaves of Grass*. In section 20 of "Song of Myself," the poet poses a question about the relationship between beef and strength:

> Who goes there? hankering, gross, mystical, nude;
> How is it I extract strength from the beef I eat?

[78] Ibid., 201.
[79] Ibid., 226.
[80] Ibid., 307.
[81] Ibid., 245.

The framing of the question, however, already circumscribes its possible answer. The question is not, "*Do I* extract strength from the beef I eat?" but instead "*How is it…?*" Whatever the answer is, it presupposes and necessarily explains the connection between strength and eating beef. Whitman's use of the verb "extract," moreover, was significant. Although Liebig did not market his Extract until the 1860s, beef teas and beef extracts had become an omnipresent medicine in America by the 1840s.[82] One can surmise that Whitman's contemporaries pondered questions similar to the poet's.

Nonetheless, not all of Whitman's signed material followed the same path. While the writer of MH&T was quite anxious and concerned about the stomach's (in)ability to digest a variety of foods, the poet of *Leaves of Grass* lacked any such anxiety. The poet of *Leaves* constantly alluded to and praised his ability to absorb and digest anything and everything in the cosmos. Whitman writes in "Song of Myself," for instance, that "In me … not a person or object missing, / Absorbing all to myself and for this song."[83] In the poem "Out of the Cradle Endlessly Rocking," the speaker describes the poetic process as "peering, absorbing, translating"[84] the world around him. And even as Whitman the poet was absorbing the world, he hoped that the favor would be returned. Indeed, for this reason he concluded the 1855 Preface to *Leaves of Grass* with his thesis that "the proof of a poet is that his country absorbs him as affectionately as he has absorbed it."[85] Unlike his dyspeptic American contemporaries, the poet—"Walt Whitman, a kosmos"—could absorb anything in combination with anything else. The poet's unflappable powers of digestion and incorporation mimic American democracy's supposed ability to absorb and assimilate new citizens of any variety. If the lethargic and enervated nations of old Europe could not tolerate foreign bodies, youthful America received and incorporated all whom it received with equal vigor. The poet even believed that his poems could be absorbed like food to strengthen the American polity: "estimating the American Union as so far, and for some time to come, in its yet

[82] Finlay, "Quackery and Cookery"; Finlay, "Early Marketing and the Theory of Nutrition: The Science and Culture of Liebig's Extract of Meat"; and Rosenberg, "And Heal the Sick."

[83] Whitman, *Poetry and Prose*, 199.

[84] Ibid., 389.

[85] Ibid., 26.

formative condition, I bequeath poems and essays as nutriment."[86] The poems and essays, Whitman hoped, would provide exemplary personalities for Americans to model and assimilate.

In one significant poem from *Leaves*, "This Compost," Whitman again drew on Liebig and the chemist's notion of organic life cycles. In it the poet questions how the earth, which continually receives sick or decaying bodies into it, nevertheless produces and sustains new life:

> O how can it be that the ground itself does not sicken?
> How can you be alive you growths of spring?
> How can you furnish health you blood of herbs, roots, orchards, grain?
> Are they not continually putting distemper'd corpses within you?
> Is not every continent work'd over and over with the sour dead?[87]

The poet expresses mystification at the Earth's immunity to disease and ability to cure or sanitize dead or decaying bodies. When the poet answers the questions posed about the ability of the earth to defy disease and death, he exclaims in wonder, borrowing the phrasing of his review of Liebig's *Chemistry in Its Application*, "Behold this compost! behold it well! ... What chemistry!"[88]

Although I know of no equivalent evidence that Whitman read Liebig's *Familiar Letters* in addition to *Chemistry in Its Application*, in the third edition of the *Letters*, published in 1851, a plausible inspiration for the third line of "Song of Myself" ("Every atom belonging to me as good belongs to you") can be found:

> Death, followed by the dissolution of the dead generation, is the source of life for a new one. The same atom of carbon, which as a constituent of a muscular fibre in the heart of a man assists to propel blood through his frame, was, perhaps, a constituent in the heart of one of his ancestors; and any atom of nitrogen in our brain has, perhaps, been a part of the brain of an Egyptian or of a negro.[89]

[86] Ibid., 1033.

[87] Ibid., 495.

[88] Ibid., 496.

[89] Liebig, *Familiar Letters on Chemistry, in Its Relation to Physiology, Dietetics, Agriculture, Commerce, and Political Economy*, 181.

What Whitman may have learned from Liebig's text enabled him to place all matter, whether organic or inorganic, on the same ontological plane.

Whitman also incorporated elements of Liebig's writings into his prose works. In *Democratic Vistas*, Whitman praises American democracy for its ability to absorb almost anything. Indeed, he explicitly compared American democracy's power to transform old social structures with nature's power to convert dead or decaying matter into nutriment for living plants and animals. In a typically complex sentence, he wrote:

> And as, by virtue of its kosmical, antiseptic power, Nature's stomach is fully strong enough not only to digest the morbific matter always presented, not to be turn'd aside, and perhaps, indeed, intuitively gravitating thither—but even to change such contributions into nutriment for highest use and life—so American democracy's.[90]

Extending the metaphor of the body politic beyond its conventional bounds, Whitman ascribed to democratic practices and institutions the ability to harness and transform any matter, sentiment, emotion, or behavior. *Democracies digest.* They, too, drew on a "kosmical, antiseptic" power that converted the old, decaying institutions of aristocracy and monarchy into nutriment for something new. Whitman understood democracy not as a set of state institutions but as something that "goes deeper." Democracy for Whitman required the abolition and transformation of old feudal hierarchies and structures, and in this sense, democracy digested the "morbific matter" of old Europe into the nutriment of a new political order.

Digestion was not therefore just relevant to politics in Whitman's texts; it constituted political activity in itself. Healthy bodies required proper diet and successful digestion, and democracy in turn required a host of those healthy bodies in order to endure as a social and political system. Drawing on nutritionist ideas, Whitman aimed to improve the health of the American (male) population, but drawing on humoral principles, he linked health and proper digestion with individual virtue. Indigestion—"the great American evil"—threatened the American revolutionary projects at its bodily foundations.

[90] Ibid., 973.

Bibliography

Brock, W. H. *Justus von Liebig: The Chemical Gatekeeper*. Cambridge, UK: Cambridge University Press, 2002.

Coleman, William. "Health and Hygiene in the Encyclopédie: A Medical Doctrine for the Bourgeoisie." *Journal of the History of Medicine and Allied Sciences* 29, no. 4 (1974): 399–421. https://doi.org/10.1093/jhmas/xxix.4.399.

Cullather, Nick. "The Foreign Policy of the Calorie." *The American Historical Review* 112, no. 2 (April 1, 2007): 337–364. https://doi.org/10.1086/ahr.112.2.337.

Finlay, Mark R. "Early Marketing and the Theory of Nutrition: The Science and Culture of Liebig's Extract of Meat." In *The Science and Culture of Nutrition, 1840–1940*, edited by Harmke Kamminga and Andrew Cunningham, 48–74. Amsterdam and Atlanta, GA: Rodopi, 1995.

———. "Quackery and Cookery: Justus von Liebig's Extract of Meat and the Theory of Nutrition in the Victorian Age." *Bulletin of the History of Medicine* 66, no. 3 (Fall 1992): 404–420.

Foucault, Michel. *The Archaeology of Knowledge; And, the Discourse on Language*. New York: Pantheon Books, 1972.

———. *The History of Sexuality. Vol. 2: The Uses of Pleasure*. Translated by Robert Hurley. New York: Vintage Books, 1990.

———. *Security, Territory, Population: Lectures at the Collège de France 1977–1978*. New York: Picador, 2009.

Graham, Sylvester. *A Lecture to Young Men on Chastity: Intended Also for the Serious Consideration of Parents and Guardians*. G.W. Light, 1838.

Kant, Immanuel. *The Conflict of the Faculties=Der Streit Der Fakultäten*. Janus Series 3. New York: Abaris Books, 1979.

Levine, Susan. *School Lunch Politics: The Surprising History of America's Favorite Welfare Program*. Princeton: Princeton University Press, 2010.

Rabinbach, Anson. *The Human Motor: Energy, Fatigue, and the Origins of Modernity*. Berkeley: University of California Press, 1992.

Rather, L. J. "The 'Six Things Non-Natural': A Note on the Origins and Fate of a Doctrine and a Phrase." *Clio Medica* 3 (1968): 337–347.

Rebrovick, Tripp. "The Politics of Diet: 'Eco-Dietetics,' Neoliberalism, and the History of Dietetic Discourses." *Political Research Quarterly* 68, no. 4 (December 1, 2015): 678–689. https://doi.org/10.1177/1065912915605183.

Rosenberg, Charles E. "And Heal the Sick: The Hospital and the Patient in the 19th Century America." *Journal of Social History* 10, no. 4 (1977): 428–447.

Scrinis, Gyorgy. "On the Ideology of Nutritionism." *Gastronomica* 8, no. 1 (February 2008): 39–48. https://doi.org/10.1525/gfc.2008.8.1.39.

———. *Nutritionism: The Science and Politics of Dietary Advice*. Arts and Traditions of the Table: Perspectives on Culinary History. New York: Columbia University Press, 2013.
Shapin, Steven. "How to Eat Like a Gentleman: Dietetics and Ethics in Early Modern England." In *Right Living: An Anglo-American Tradition of Self-Help Medicine and Hygiene*, edited by Charles E. Rosenberg, 21–58. Baltimore: Johns Hopkins University Press, 2003.
———. "'You Are What You Eat': Historical Changes in Ideas About Food and Identity." *Historical Research* 87, no. 237 (August 2014): 377–392. https://doi.org/10.1111/1468-2281.12059.
Siraisi, Nancy G. *Medieval & Early Renaissance Medicine: An Introduction to Knowledge and Practice*. Chicago: University of Chicago Press, 1990.
Taylor, Charles. "Foucault on Freedom and Truth." *Political Theory* 12, no. 2 (May 1, 1984): 152–183. https://doi.org/10.1177/0090591784012002002.
Temkin, Owsei. *Galenism: Rise and Decline of a Medical Philosophy*. Cornell Publications in the History of Science. Ithaca: Cornell University Press, 1973.
Turpin, Zachary. "Introduction to Walt Whitman's 'Manly Health and Training.'" *Walt Whitman Quarterly Review* 33, no. 3 (April 29, 2016): 147–183. https://doi.org/10.13008/0737-0679.2205.
von Liebig, Justus. *Chemistry in Its Application to Agriculture and Physiology*. Edited by Lyon Playfair Playfair. J. Owen, 1842.
———. *Familiar Letters on Chemistry and Its Relation to Commerce, Physiology, and Agriculture*. Edited by John Gardner. 1st ed. New York: D. Appleton & Co., 1843.
———. *Familiar Letters on Chemistry, in Its Relation to Physiology, Dietetics, Agriculture, Commerce, and Political Economy*. Edited by John Blyth. 3rd ed. London: Taylor, Walton, & Maberly, 1851.
Whitman, Walt. "Review of Liebig's Chemistry in Its Application to Agriculture and Physiology." *Brooklyn Daily Eagle*, 28 June 1847.
———. *Poetry and Prose*. 1st Library of America College Editions. New York: Library of America, 1996.
———. *The Correspondence*. Edited by Ted Genoways. Vol. 7. Iowa City: University of Iowa Press, 2004.
———. *Walt Whitman's Songs of Male Intimacy and love: "Live Oak, with Moss" and "Calamus."* Edited by Betsy Erkkila. The Iowa Whitman Series. Iowa City: University of Iowa Press, 2011.
———. "Manly Health and Training, With Off-Hand Hints Toward Their Conditions." *Walt Whitman Quarterly Review* 33, no. 3 (April 29, 2016): 184–310. https://doi.org/10.13008/0737-0679.2206.

CHAPTER 3

The "Second Brain": Dietetics and Ideology in Nineteenth-Century France

Bertrand Marquer

In a famous passage from Zola's *L'Œuvre* (*The Masterpiece*, 1886), the character named Sandoz sums up the Naturalist credo and in doing so calls for the organic hierarchy to be reworked so as to give physiology the place it deserves:

> Don't you think it's a farce, the way they've concentrated their studies exclusively on the brain and its functions and pretend that the brain is the noblest of the human organs? … Thought, now. What is thought, in God's name, but the product of the entire body? Can they get a brain to think by itself? What happens to the 'nobility' of the brain when its owner has belly-ache?…[1]

[1] Zola, *The Masterpiece*, 180. ["*N'est-ce pas une farce que cette étude continue et exclusive de la fonction du cerveau, sous le prétexte que le cerveau est l'organe noble? La pensée, la pensée, eh! tonnerre de Dieu! la pensée est le produit du corps entier. Faites donc penser un cerveau tout seul, voyez donc ce que devient la noblesse du cerveau, quand le ventre est malade!*"]

B. Marquer (✉)
University of Strasbourg and Institut Universitaire de France, Strasbourg, France

M. Mathias and A. M. Moore (eds.), *Gut Feeling and Digestive Health in Nineteenth-Century Literature, History and Culture*, Palgrave Studies in Literature, Science and Medicine,
https://doi.org/10.1007/978-3-030-01857-3_3

This provocative observation is reminiscent of the one that the French physiologist Pierre-Jean-Georges Cabanis offered in his thesis, *Rapports du physique et du moral* (*Reports on the Physical and the Moral Aspects of Man*, 1802). Cabanis claimed that "the brain digests impressions and organically secretes thought", drawing an analogy between the stomach and the brain that worked in both directions. The "physiology of thought" formulated by Zola's Sandoz is thus far from being an isolated case, and it calls to mind both the medical and gastronomic discourses of the first half of the nineteenth century in France.

A poem published in 1830 in *Le Gastronome*,[2] for example—"La Chanson à digérer" ("A Song to be digested")—states that physiologists have discovered, "through a new system", that "our moods are made happy or sad by the stomach and not by the brain".[3] The anonymous author[4] then concludes that "like style", "the belly is the whole man": "Digest well, and you will become perfect!".[5] In both cases, the references to the stomach and the belly assert the brain's dependence on the physiology of digestion. The image of the second brain reappears in Honoré de Balzac's novel, *Le Cousin Pons* (1847) with the notion of the gut as a "second brain, placed in the diaphragm".[6] It is the symbolism of this "second brain" that will be analysed in this essay, revealing the functions and meanings of this physiological imaginary in nineteenth-century France.

In the post-revolutionary period, the physiology of digestion was at the centre of a vast network of texts, the nature of which were often very different. Their common ground was in making a physiological imaginary embodying the new balance between body and thought established

[2] The full title of this journal is *Le Gastronome, journal universel du goût, rédigé par une société d'hommes de bouche et d'hommes de lettres* (*The Gastronome, Universal Journal of Taste, Written by a Society of Men of the Table and Men of Letters*).

[3] "*Mangeant, buvant, les physiologistes / Ont découvert, en système nouveau, / Que nos humeurs sont joyeuses ou tristes / Par l'estomac et non par le cerveau.*"

[4] The author is likely to be Le Bibliophile Jacob, the alias of the journalist and polygraph, Paul Lacroix.

[5] "*De là dépend tout le bonheur: en somme, / Petite cause amène grands effets! / Comme le style, ah! le ventre est tout l'homme… / Digérez bien, pour devenir parfaits*". In French, the penultimate verse ("*Comme le style, ah! le ventre est tout l'homme*") is a transparent allusion to Buffon's famous statement, pronounced in 1753 at l'Académie Française: "Style is the man himself" ("*Le style est l'homme même*").

[6] Balzac, *Le Cousin Pons*, 36.

at the turn of the century by ideologists like Cabanis, and then widely communicated, in the early nineteenth century, by means of a gastronomic discourse anxious to propose new reflections on man in general. The rise of gastronomy in the nineteenth century was closely linked to the political reconfiguration engendered by the French Revolution. Presented as a means of rebuilding the social fabric, gastronomy was a key element in the political affirmation of the bourgeoisie: it symbolized a transition between aristocracy and oligarchy, while appearing as a creation of the new society that resulted from the Revolution, acting as an allegory of a new order. Gastronomy in the first third of the nineteenth century commonly evoked the "power of this cosmopolitan sovereign"[7]: for the author of *Le Code gourmand* ("The Law of Gluttony"), a work inspired both by Brillat-Savarin and by Napoléon's *Code civil*, gastronomy was indeed "the queen of the world" (ibid.), a queen that provided a set of social and practical rules.

Physiology played a central role in the construction of new gastronomic norms, asserting the preponderant role of the stomach as a regulating organ and the guarantor of health. The image used by the doctor Théophile de Bordeu, for example, of the "triumvirate, tripod of life"[8]—incorporating the stomach, the heart and the brain—was repeated in popularizations of science in this period,[9] and early in the century, it was used to reinforce the physiological importance of the stomach to the detriment of the other two vital centres, the heart and the brain.[10] The physiology of digestion at this time thus tends to confirm the ancient parable of the Belly and the Members. This political parable, adapted by La Fontaine in his *Fables* (1694), made the stomach the centre of the social body, and showed that a revolt of the members (tired of working for an organ deemed unnecessary), signified the death of the body as a whole. With the revolutionary upheavals of the turn of the century, this parable acquired a new significance and was subsequently reconfigured:

[7] Raisson, *Code gourmand*, 5.

[8] Bordeu, *Recherches sur les maladies chroniques*, 73.

[9] See in particular the works written by the doctor and historian of medicine Joseph-Michel Guardia, for whom Bordeu "embodied the medicine of the eighteenth century and prepares that of the nineteenth" ["*C'est en Théophile de Bordeu que s'incarne, pour ainsi dire, la médecine française du XVIIIe siècle, [...] qu'émane en grande partie celle du XIXe*"] (1887, 127).

[10] Marquer, "De l'épigastre au ventre."

if the head was associated with the monarchy, the post-revolutionary French society made of the stomach a major actor of the social body, and affirmed the new political centrality of this physiological intermediary.

Closely related to political representations, the physiology of the stomach was also a privileged vector of ideology, as is confirmed by its central place in the hygienist discourses of the early nineteenth century. More broadly, contemporary medical discourse on the stomach supported the political changes underway, and contributed to the construction of a new norm, through "scientific ideologies" (Canguilhem, 1977)[11] relayed throughout the century by a physiology of digestion. This physiology was sometimes considered as an instrument of social regulation (as in the case of social Darwinist theories), and sometimes as a matrix of thought (the stomach as a "second brain"). In both cases, the physiological analogy legitimizes the shift from the medical standard to the moral norm, and, as we shall see, makes the stomach the symbolic pivot for a system of values that had a very wide field of application.

Dietetics and the Physiology of the Brain

The first introduction to this physiological imaginary was provided within the country's gastronomic discourse, which developed in the wake of Grimod de la Reynière's writings on food and dining such as *L'Almanach des gourmands* (1803–1813), and especially Brillat-Savarin's best-selling *Physiologie du goût* (1825) [*Physiology of taste*]. As was the case for Cabanis, the main aim of these gastronomic texts was to promote a materialist ideology, in both strategic and playful ways. Menenius Agrippa and his famous parable of the Belly and the Members were thus mentioned in the gastronomic press to confirm the high functions of the stomach. In an article called "Les caprices de l'estomac" ("The Whims of the Stomach"), published in *La Gastronomie* in 1840, the "immortal fable" of the Roman consul serves as a starting point for singing the stomach's praises. According to the mysterious author of this article, M. de V., the stomach was "the focal point of animal life" and

[11] According to the definition of the French philosopher and physician, Canguilhem, scientific ideologies are "explanatory systems whose object is hyperbolic relative to the standard of scientificity they borrow" (*Idéologie et rationalité dans les sciences de la vie*, 44).

"the general collector of the human body".[12] If "the head is the seat of the will, intelligence, and other dubious sentimentalities, it is none the less the stomach's humble servant; at its pleasure, the stomach keeps the headlight or makes it heavy and makes it bear the punishment of its sins".[13] The long "Dialogue philosophico-médico-gastronomico-littéraire" between the brain and the stomach invented by Le Bibliophile Jacob took the same direction. The author of this dialogue puts the Cartesian cogito on the same plane as digestion: "I think: therefore I exist, said in itself a brain full of the meditations of Descartes"; "I digest: so I live, whispered a voice from below".[14] Finally, Le Bibliophile Jacob even asserts the superiority of the inhabitant of the "ground floor" (the stomach) over "Lord Cervelet" (Lord Cerebellum). The depiction of the anatomical hierarchy comically takes its model from the nineteenth-century social hierarchy as reflected in the class layers of a Parisian apartment building, in order to rehabilitate the digestive functions and the influence of the stomach on the brain: since they live on "the first floor", the "Eaters" appear as the superiors of the "Thinkers" who are "up in the garret". By sending a "colic" to the brain, the stomach also forces it to relativize its omnipotence, and finally to admit that they must be "treated as equals": "I am as much as you, and I would like to persuade you, once and for all, that my face is worthy of yours, that my functions are more extensive, more important than you think, and that your products, without mine, will never be anything but hollow dreams".[15]

Although the article is written with ironic intent, another article on "Les gros ventres" ("Big Bellies", 1830) leads to the same conclusion: on the basis of the "acknowledged superiority of what are called 'les gros

[12]"*L'estomac, foyer de la vie animale, est le laboratoire où la digestion, reine toute-puissante, envoie à chacun de ses ministres la portion de vie qui concerne son département, vérité admirablement formulée, dès les premiers temps de Rome, par Menenius Agrippa, dans son immortelle fable:* Les Membres et l'Estomac. *Le fait est que l'estomac est le receveur général du corps humain*" (M. de V., 1840, 2).

[13]"*la tête est le siège de la volonté, de l'intelligence et autres mièvretés fort douteuses, la tête n'en est pas moins la très-humble servante de l'estomac; à son gré, il la maintient légère ou la rend pesante; il lui fait supporter le châtiment de ses fautes*" (ibid.).

[14]Le Bibliophile Jacob, "Les gros ventres," 1830a.

[15]"*Maintenant, traitons d'égal à égal; car je suis autant que toi, et je voudrais te persuader, une fois pour toutes, que ma figure vaut la tienne, que mes fonctions sont plus étendues, plus importantes que tu ne crois, et que tes produits, sans les miens, ne seront jamais que des rêves creux*" (Bibliophile Jacob, "Les gros ventres," 1830a, 2).

messieurs'", the author takes up "the idea of establishing the seat of the human soul in the abdominal viscera" in order to envisage a "gastrology" modelled on the "craniology of Doctor Gall", since the "conformation [of Messire Gaster], and its capacity, are the right measure of our intelligence".[16] By referring to the allegorical character of Rabelais' *Quart Livre* (Messire Gaster), Le Bibliophile Jacob confirms this symbolic equivalence between the brain and the stomach which had become the standard by which to measure each person's intelligence.

On the whole, physiological reflections about the influence of diet on the brain were treated with humour by the gastronomic press, stressing the need, for example, for fasting, or the impact of a stimulating or soothing diet. In this case, the link between the brain and the stomach was all the easier to establish, since the dietary discourse was largely based on the theory of humoral temperaments. In a book entitled *Le Cuisinier et le médecin* (*The Cook and the Doctor*, 1855), the French physician Léandre-Moïse Lombard, for instance, uses the "six species of temperament referred to by Cabanis and Richerand"[17] to establish a typology of diets ("refreshing", "relaxing", "tonic" or "aphrodisiac"[18]). The essential thing, for Lombard, was "to make the activity of the brain free, in accordance with that of the stomach", since "instead of harming one another, they help each other. A quick and easy digestion favours the lucidity and penetration of intelligence".[19] In his "critical encyclopedia

[16] "*De cette supériorité reconnue dans ce qu'on appelle les gros messieurs naquit sans doute l'idée d'établir le siège de l'âme humaine dans les viscères abdominaux, rêverie ingénieuse qui rendrait messire Gaster responsable de toutes nos facultés: sa conformation, sa capacité offriraient alors la juste mesure de notre intelligence. […] Nous avions déjà la crânologie du docteur Gall: pourquoi la gastrologie n'aurait-elle pas son tour?*" (Bibliophile Jacob, "Les gros ventres," 1830b, 5).

[17] Lombard, *Le Cuisinier et le médecin*, 6.

[18] "Refreshing food" ("*l'alimentation rafraîchissante*") is composed of "foods that contain a slightly acidic principle" ("*aliments qui contiennent un principe légèrement acide*") like vegetables or berries (Lombard, 1855, 44). The "releasing diet" ("*alimentation relâchante*") is composed of fatty foods, and produces the "lymphatic temperament" ("*tempérament lymphatique*," 45). "Tonic food" is based on "vegetable and bitter substances" ("*substances végétales et amères*," 46).

[19] "*au lieu de se nuire, l'un aide à l'autre [sic], et […] une digestion prompte et facile ajoute à la lucidité et à la pénétration de l'intelligence*" (Lombard, *Le Cuisinier et le médecin*, 117).

of farce",[20] Flaubert similarly devoted a note to the "influence of the digestive organs", concluding that "digestion exerts on the intellect the most frequent and often the most powerful influence": "everyone thinks according to his brain, as everyone digests according to his stomach" (Flaubert, n.d.).[21] It was therefore not only in gastronomic texts that the power of the stomach over the brain was explored, but also in literary works.

It is not surprising, therefore, to see a poet prefacing a medical work on the hygiene of the stomach, in order to emphasize his high opinion of such an important subject. Théodore de Banville praised the "admirable book" written by Doctor Monin, "one of the *coryphée* ('leading lights')[22] of hygienic popularization", by stating[23]:

> Read Doctor Monin, weigh his advice well, and if possible, try to follow it. For the hygiene of the stomach is also the hygiene of the mind and that of the soul: it is courage, it is the clarity of thought, it is virtue, it is bravery![24]

According to Banville, the "sympathetic relations" between the intellectual and digestive functions mentioned by Doctor Lombard (1855, 134) are obvious. They presuppose—or even impose—dietary precautions, in order to guarantee "clarity of thought" and other moral virtues.

Lombard, when discussing the "diet of men of letters", also drew attention to the danger posed by "intestinal neurosis".[25] Monin, meanwhile, provided physiological evidence to support Brillat-Savarin's famous aphorism ("Tell me what you eat, I will tell you what you are"[26]), by pointing out that "everyone has his gastric individuality".[27]

[20] Flaubert, "Lettre à Edma Roger des Genettes," 559.

[21] Gustave Flaubert, n.d., "Dossier G226—Vol. 7."

[22] The coryphée is the leading dancer of a corps de ballet.

[23] Favre, *Dictionnaire universel de cuisine et d'hygiène alimentaire*, XIV.

[24] "*Lisez le docteur Monin, pesez bien ses conseils, et s'il se peut, tâchez de les suivre. Car l'hygiène de l'estomac, c'est aussi l'hygiène de l'esprit et celle de l'âme: c'est le courage, c'est la clarté de la pensée, c'est la vertu, c'est la bravoure!*" (n.d., VIII).

[25] Lombard, *Le Cuisinier et le médecin*, 127.

[26] "*IV. Dis-moi ce que tu manges, je te dirai ce que tu es*" (Brillat-Savarin, *Physiologie du goût*, 12).

[27] "*Chacun a, si l'on peut dire, son individualité gastrique*" (Favre, *Dictionnaire universel de cuisine et d'hygiène alimentaire*, V).

The "extensive influence" of digestion "on ideas, on passions, on habits, in a word, on one's moral state"[28] mentioned by Cabanis not only resulted in the development or atrophy of intellectual functions, but actually gives birth to a true physiology of thought in which food is not merely a reflection of the consumer's taste, but a determinant of personality.

"The Imagination of the Stomach": Dietetics and Ideology

This influence of digestion on thought also extended to artistic production, contributing not only to its content, but also to the poetic forms chosen, as Edmond de Goncourt noted in his *Journal*: "it would be interesting for an intelligent author to write several books of imagination: one under the influence of a coffee diet, another one under the influence of a tea diet, another one of a wine and alcohol diet. He would be able to study the influences of these stimulants on his writing and he would share them with the public".[29] In his own words, Edmond de Goncourt "believes that food has a great influence on literary production" and he even plans, "one of these days", to undertake the same work (such as "a short story or an act") twice: once with "restricted food and washed down with plenty of tea", and once with "very powerful foods and a lot of coffee".[30] These remarks by Goncourt should not be read as a simple joke: they reflect the extent of this belief in the influence of diet on the productions of the mind, of which literary creation is a part.

The "imagination of the stomach" which Goncourt evoked several times in his *Journal*,[31] thus referred not only to the culinary art of the

[28] Cabanis, *Rapports du physique et du moral*, II, 133.

[29] "*Il serait intéressant qu'un littérateur intelligent fît plusieurs livres d'imagination: l'un au régime du café, l'autre au régime du thé, l'autre au régime du vin et de l'alcool, et qu'il étudiât sur lui les influences de ces excitants sur sa littérature et qu'il en fît part au public…*" (Goncourt, *Journal. Mémoires de la vie littéraire*, vol. 3, 279, written June 9, 1889).

[30] "*Je crois que la nourriture a une grande action sur la production littéraire. À défaut du même travail – si l'on pouvait oublier –, que je voudrais deux fois faire avec deux alimentations diverses, un de ces jours, je tenterais de faire une nouvelle ou un acte, avec une nourriture restreinte et lavée de beaucoup de thé, et un autre acte ou nouvelle, avec une nourriture très puissante et beaucoup de café*" (*Journal. Mémoires de la vie littéraire*, vol. 3, 866, written August 24, 1893).

[31] Goncourt, *Journal. Mémoires de la vie littéraire*. vol. 2, 929, 1130.

gastronome but rather, it conveyed the close relationship between the life of the stomach and the life of the soul. In the second half of the century, the link between digestion and creation therefore goes well beyond the influence of food on the reproductive system, mentioned, for example, in Balzac's *Traité des excitants modernes* (*Treatise on Modern Stimulants*, 1839). In this text, Balzac had argued that the first "gastronomic meditation" devoted to the senses—Brillat-Savarin's *Physiologie du goût* (*Physiology of Taste*, 1825)—was incomplete, and Balzac thus developed the analysis further by explaining that "food is generation".[32] In the later nineteenth century, however, the stomach acquired the dignity of a "second brain", and in doing so, the stomach suffered all the consequences of hosting "*la folle du logis*". In French, this image refers to the imagination, compared with an insane woman who jostles the domestic order ("*le logis*", home) of life in the brain.

This image of the imagination being located in the so-called second brain was repeatedly emphasized in the hygiene treaties of this period. For example, as early as the beginning of the century, Esquirol's nosography of mental diseases situated the passions and their pathological zone of influence "in the epigastric region".[33] The famous psychiatrist pointed out that the "melancholic man" was characterized by the displacement of the transverse colon,[34] a lesion that provided a seat for the ancient "black bile", a "captive and morose" substance which critic Jean Starobinski sees as the "image of a constrained interiority".[35] Melancholy was therefore associated with digestive disorder as much as with a psychological disease, both characterized by the formation of a blockage. "The obvious alteration of the epigastric organs"[36] even became a distinctive sign of monomaniacs, anticipating the association of the melancholic man with the dyspeptic one, both brought together at the end of the century under the pathological banner of the "slowing down of nutrition".[37] As a result, dyspepsia inherited the symbolism associated with melancholy and tended to become the hallmark of the man of genius. Designated as "the

[32] "L'alimentation est la génération" (Brillat-Savarin, *Physiologie du goût*, 452).

[33] Gauchet and Swain, *La Pratique de l'esprit humain*, 330.

[34] Esquirol, *Des maladies mentales*, vol. 1, 445.

[35] Starobinski, "L'encre de la mélancolie," 411.

[36] Esquirol, *Des passions considérées comme causes*, 18.

[37] "*les ralentis de la nutrition*," Fleury, *Introduction à la médecine de l'esprit*, 263

scholar's illness",[38] dyspepsia was considered in the nineteenth century as a disease typically found in the man of letters. Reflections on this disease built on the best-selling book published by the Swiss physician Samuel-Auguste Tissot, *De la santé des gens de lettres* (1766), whose axioms were taken up in the medical literature of the nineteenth century. Referring to the health of men of letters, Tissot asserted that "the man who thinks the most, is the one who digests the worst; all things being equal, whoever thinks the least, is the one who digests the best".[39]

This association between the scholar and dyspepsia also highlighted the impact of social imagery in this physiological conception of thought, modelled on (good) digestion: hygienic considerations based on dietetics could reactivate or reinforce social representations and contribute to the stigmatization of certain lifestyles, as in the case of the writer. This is evidenced by the "Gendelettra morbus" invented by a certain "Colin Tampon" for the *Figaro* literary supplement (1890), a humorous response to a chronicle on dyspepsia written by Dr. Maurice de Fleury. The journalist imagined, on the model of cholera, a "terrible epizootic" that decimated men of letters, designated as "*bêtes à plumes*": "feathered beasts" (in French, there is a play on words between the pen nib and the word for "feather"). The situation required "strong measures" to curb this "scourge" that threatened literature itself. This ironic observation then allows Colin Tampon to joke about the danger of the writing profession, arguing that medical statistics should now rank literature among the unhealthy professions, alongside the most dangerous manual trades. The traditional discredit attached to the profession of the writer is put into perspective here: if, "formerly", respectable families warned their offspring against "the risk of dying of hunger", this fate was less cruel than that which doctors now foresaw for the men of letters: "succumbing to the practice of fasting, drinking chamomile, with a novel from which twenty-five editions have been printed, or a play which has passed its hundredth performance. Sweet mothers", concludes Colin Tampon, "tremble for your offspring and save them from that frightful disease that rises from the swamps of the inkwell!".[40]

[38] "*la maladie des gens de lettres*," Seure, 1885, 251.

[39] See, e.g., Seure, *Dyspepsie et dyspeptiques*, 251. The quotation comes from Tissot, *De la santé des gens de lettres*, 32.

[40] "*Sans doute, de tout temps, les familles ont vu d'un mauvais œil leurs rejetons s'engager dans une carrière où il était fréquent que l'on mourût de faim, mais mourir de faim à la*

Threatened with melancholy and dyspepsia, the nineteenth-century man of letters was now considered to be undertaking a highly risky form of employment. The hygienist discourse of the period therefore confirmed contemporary social doxa on the marginality of the artist: since a man's diet did not only condition his physiognomy, but also his psychology, it supported a norm that went far beyond the physiological field. In his chapter devoted to "Rules of hygiene and variability of digestion according to professions", for instance, Dr. Lombard provided a surprising elaboration on the notion of regime as defined by Cabanis, attributing political connotations to the moral effects noted by the author of *Rapports du physique et du moral*[41]:

> More robust and healthier in mind and body, the hardworking bourgeoisie lives next to and above the working class; [...] this labouring part of our urban populations is well-off and strives by work, by order, by habits of economy, by the simplicity of manners, to attain independence. It is from this class that the men who are at the head of the country are recruited; it is this class that has furnished most of the serious minds which, for sixty years, have honoured the nation. It is also in this portion of society that good manners, regularity, a carefully selected diet, the contentment of the soul, and moderate exercise of body and mind, maintain easy digestion and, consequently, vigour and health.[42]

manière d'autrefois c'était peut-être moins cruel que de succomber à la diète, en buvant de la camomille, avec un roman dont on a tiré vingt-cinq éditions ou une pièce de théâtre qui a franchi sa Centième! Tendres mères, tremblez pour votre progéniture et sauvez-là de cette affreuse maladie qui s'élève des marécages de l'encrier" (Colin Tampon, "Gendelettra morbus").

[41] See Cabanis, *Rapports du physique et du moral*, 133: these "moral effects" are related to the quality and quantity of food consumed, and are considered from a strictly dietetic point of view.

[42] "*Plus robuste, plus saine d'esprit et de corps, vit à côté et au-dessus de la classe ouvrière, la bourgeoisie travailleuse, celle qui exerce dans nos grandes villes le commerce de détail et l'industrie; cette partie laborieuse de nos populations urbaines possède l'aisance et s'efforce par le travail, par l'ordre, par les habitudes d'économie, par la simplicité des mœurs, d'arriver à l'indépendance. C'est dans cette classe que se recrutent les hommes d'élite qui marchent à la tête du pays; c'est elle qui a fourni la plupart des esprits sérieux qui, depuis soixante ans, l'ont illustré. C'est aussi dans cette portion de la société que les bonnes mœurs, la vie régulière, l'alimentation choisie, le contentement de l'âme, l'exercice modéré du corps et de l'intelligence, maintiennent les digestions faciles et par suite la vigueur et la santé*" (Lombard, *Le Cuisinier et le médecin*, 112).

Dietetics therefore also had axiological consequences, by confirming and legitimizing a social hierarchy. The bourgeois regime was indeed associated with a prescriptive hygiene and a philosophy of the *aurea medocritas* (Horace, odes, II, 10), a way of life based on temperance and a happy medium. It seemed to provide the physiology of digestion with a norm that embraced the biological and social body, showing how the "mastery of digestion" participated in the "scientific and cultural ambition" of a social class.[43] The development of gastronomy, as well as the emergence of the restaurant as a new symbolic place of power,[44] both contributed to the definition of a bourgeois ethos in which the way of eating (a "*savoir-manger*") became the counterpart of the traditional good manners (the "*savoir-vivre*") of the nobility. The imitation of the commensal habits and customs of ancient aristocracies was a process that anticipated and accompanied the adoption of political functions by the bourgeoisie in the nineteenth century.[45] In the self-representation that this social class adopted, gastronomy constituted a "prestige value", "the sign of its social promotion",[46] and it therefore became a key way of measuring the symbolic recovery of noble values.

Socially connotated, this "mastery of digestion" also provided arguments and images that confirmed the "scientific ideologies" repeated or discussed in literary texts. By supporting a cultural and physiological standard of good manners (a "s*avoir-vivre*"), the way of digesting (a "*savoir-digérer*") participated, for example, in the definition of a deviance that included both the dyspeptic artist and the voracious proletarian. The famous meal given by Gervaise in Zola's best-selling *L'Assommoir* (1877) is emblematic in this respect. The guests clumsily imitate the codes of the bourgeois meal with the desire to "stuff themselves" ("*s'en coller une bosse*"), and end up rendering, literally, what they have consumed:

> You could see their bellies swell up while they ate, you really could. The ladies looked in the family way. Every one of 'em bursting out their skins, the bleeding hogs! With their mouths open and their chins dripping with

[43] Vigarello, *Les Métamorphoses du gras*, 205.

[44] See Aron, *Le Mangeur du dix-neuvième siècle*.

[45] Albert, *Aux tables du pouvoir*, 180–181.

[46] Aron, *Le Mangeur du dix-neuvième siècle*, 13.

grease, their faces were like backsides, and were so red, you'd have said rich folk's backsides, bursting with prosperity. [...] Clemence wasn't the kind of girl you'd ask again: she ended up showing off everything she had, and she puked all over one of the muslin curtains and completely ruined it. At least the men did go into the street to do it; Lorilleux and Poisson, when they felt queer, managed to dash as far as the pork-butcher's shop. Breeding always tells.[47]

As in Bakhtin's analysis of Rabelais, the Carnival ritual is for Zola a way to signify that the inversion of the physiological cycle of digestion amounts to a questioning of the established social order. The mastery of one's own stomach symbolizes both biological and political norms, threatened by the intemperance of Gervaise and her guests. This Carnival trope of inversion manifests itself in particular in the association of the guests' faces with "rich folk's backsides".

More generally, however, the mastery of digestion symbolized and reproduced the social hierarchy, whether in "panoramic literature",[48] or in medical discourse. Eugène Briffault conducted his investigation of meals in Paris (*Paris à table*), for example, by beginning on the "summit" and then going down "each step of the social ladder", in order to demonstrate that each class is reflected by its table manners.[49] According to the Hungarian doctor Max Nordau, known for his satire of the degenerated "*fin de siècle*" France (in French in the original edition of *Entartüng*, this process was referred to as "degeneration"), the ability to digest also served as a criterion for the definition of genius. Just like the chlorophyll plant (which he thought was the only form of life capable of transforming carbon and nitrogen into nutrients), man's

[47] Zola, *L'Assommoir*, 215–216, 231–232. ["*Vrai, on voyait les bedons se gonfler à mesure. Les dames étaient grosses. Ils pétaient dans leur peau, les sacrés goinfres! La bouche ouverte, le menton barbouillé de graisse, ils avaient des faces pareilles à des derrières, et si rouges, qu'on aurait dit des derrières de gens riches, crevant de prospérité. [...] Ce que chacun déclarait peu propre, c'était la conduite de Clémence, une fille à ne pas inviter, décidément; elle avait fini par montrer tout ce qu'elle possédait, et s'était trouvée prise de mal de cœur, au point d'abîmer entièrement un des rideaux de mousseline. Les hommes, au moins, sortaient dans la rue; Lorilleux et Poisson, l'estomac dérangé, avaient filé raide jusqu'à la boutique du charcutier. Quand on a été bien élevé, ça se voit toujours.*"]

[48] Benjamin, *The Arcades Project*. Walter Benjamin describes this genre as a series of social sketches or "moral dioramas [...] of unscrupulous multiplicity" (*The Arcades Project*, 6).

[49] Briffault, *Paris à table*, 12–13.

"genius transforms, with the organs of digestion that he alone possesses, the impressions of the senses into a work of art comprehensible, digestible for other humans".[50] Dyspepsia became, under these conditions, the sign of the "non-genius [which] cannot digest nature, assimilate it, transform it into constituent parts of its own consciousness".[51] Defined by the French physician, Réveillé-Parise, as "a function which consists in the affinity of self for itself, in the attraction of similar elements",[52] digestion guaranteed the recognition of a shared talent and participated in the Bourgeois *ethos*.

The taste for corrupt food and for decomposition manifested by Decadent literature at the end of the century was therefore situated within broader ways of thinking that value temperance as a social and aesthetic model, and associate health or good taste with academism. Whether valued or stigmatized, the depravity of taste appeared as the reverse of the normalized digestion personified by the bourgeoisie. Medical diagnosis thus contained a form of aesthetic appreciation, as in Huysmans' *À rebours* (1884). The "nervous dyspepsia" suffered by Des Esseintes was part of a medico-aesthetic reading that ran through the whole of the nineteenth century, but which imposed itself with more force in the last decades when the stomach (and its health) clearly became the subject of both medical and aesthetic evaluation. Thus, according to Nordau, the famous symbolists, Jean de Moreas (the author of the Symbolist Manifesto) and Stanislas de Guaita (the founder, with Joséphin Péladan, of the Kabbalistic Order of the Rose-Cross), wielded a "bric-a-brac of erudition" in which "the names of Schopenhauer, Darwin, Taine, Renan, Shelley, Goethe" appeared like "undigested scraps in an unrecognizable garnish".[53] As for the mystic Maeterlinck, who "stuffed his stomach with Shakespeare and brought his plays up without digesting

[50] "le génie transforme, avec les organes digestifs intellectuels que lui seul possède, les impressions des sens en une œuvre d'art compréhensible, digestible pour les autres humains" (Nordau, *Psycho-physiologie du génie et du talent*, 58).

[51] "le non génie ne peut digérer la nature, se l'assimiler, la transformer en parties constitutives de sa propre conscience" (Nordau, *Psycho-physiologie du génie et du talent*, 56–57).

[52] "une fonction qui consiste dans l'affinité de *soi pour soi*, dans l'attraction des éléments similaires" (Réveillé-Parise, *Physiologie et hygiène*, I, 234.

[53] "*les noms de Schopenhauer, Darwin, Taine, Renan, Shelley, Gœthe, [...] servent d'étiquette à des rognures informes et méconnaissables, à des balayures de bribes non digérées*" (Nordau, *Dégénérescence*, 181).

them, transforming them", he transformed them into "a repugnant mass on the verge of putrid decomposition".[54] Nordau affirms that "this image is unpalatable, but it is the only way of giving a clear idea of the intellectual process that occurs as soon as the degenerates do what they call 'create'".[55]

Unsurprisingly, the criteria used by the hygienist were on the same side as the academy when applied to literature: the criticism used by Nordau, which we might refer to as dietetic criticism, aimed to reinforce the norms that structured the social imaginary. The physiological categories of health and unhealth therefore conveyed axiological judgements on the period's aesthetic productions, and even those of the nation as a whole: pica or malacia (diseases of food perversion) symbolized the depravity of aesthetic taste, just as dyspepsia stood for the disturbance of the appetite for beautiful things.

In the nineteenth century, therefore, the image of the "second brain" was part of a physiological imaginary that often served a normative social system, in which the stomach was an allegorical and metonymic organ. It was allegorical insofar as knowing how to digest was linked to the knowledge of how to live, in a manner specific to the bourgeoisie. It was metonymic in the sense that the physiology of digestion served as a model for the physiology of the brain, of thought, and ultimately of the body in general. In both cases, digestion offered a model of meaning and interpretation by borrowing or transfer: from the biological body to the social body and from the physiological to the psychological. On the aesthetic level, the new norm, which the "second brain" helped to shape, played on the ambivalence of taste itself in order to give a physiological foundation to the aesthetic canons being defended. This "dietetic" criticism of works of art ultimately shows the extent of the interactions between physiological discourse and literary discourse in the nineteenth century, which is particularly significant for our understanding of representations of the gut in this period. References and stereotypes relating to food circulated from one domain to another and interacted with each other, accentuating the connotations and symbolisms

[54] Réveillé-Parise, *Physiologie et hygiène*, I, 234.

[55] "*il s'est gavé l'estomac de Shakespeare et rend les morceaux non digérés, mais transformés d'une manière répugnante et avec un commencement de décomposition putride. Cette image n'est pas ragoûtante, mais elle peut seule donner une idée claire du processus intellectuel qui se produit lorsque des dégénérés font ce qu'ils appellent 'créer'*" (Nordau, *Dégénérescence*, 420).

related to taste. Since it was a polysemic notion, both physiological and aesthetic, taste was subject to ideological issues, and the physiology of digestion, connected to that of the brain, was used to support and communicate broader scientific and social ideologies. In this case, the definition of health therefore influenced norms that were no longer simply physiological, but also ideological. The nineteenth-century concept of the "second brain" allows us to appreciate how physiology can also be cultural.

Bibliography

Albert, Jean-Marc. *Aux tables du pouvoir. Des banquets grecs à l'Elysée*. Paris: Armand Colin, 2009.

Aron, Jean-Paul. *Le Mangeur du dix-neuvième siècle. Une folie bourgeoise: la nourriture*. Paris: Robert Laffont, 1973.

Benjamin, Walter. *The Arcades Project*. Translated by Howard Eiland and Kevin McLaughlin. Cambridge: The Harvard University Press, 1999.

[Bibliophile Jacob]. "Le cerveau et l'estomac. Dialogue philosophico-médico-gastronomico-littéraire." *Le Gastronome*, no. 2 (18 mars 1830a): 2–4.

Bibliophile Jacob. "Les gros ventres." *Le Gastronome*, no. 29 (20 juin 1830b).

Briffaut, Eugène. *Paris à table*. Paris: Hetzel, 1846.

Brillat-Savarin, Jean Anthelme. *Physiologie du goût, ou méditations de gastronomie transcendante. Ouvrage théorique, historique, et à l'ordre du jour, dédié aux gastronomes parisiens, par un professeur, membre de plusieurs Sociétés savantes, suivie d'un traité sur les excitants modernes de M. de Balzac*. Paris: Charpentier, 1839 [1825].

Cabanis, Pierre-Jean-Georges. *Rapports du physique et du moral de l'homme*, Paris: Crapart, Caille et Ravier, 1805 [1802].

Canguilhem, Georges. *Idéologie et rationalité dans les sciences de la vie*. Paris: Vrin, 1977.

de Balzac, Honoré. *Le Cousin Pons*. Paris: Gallimard, 1973 [1847].

de Bordeu, Théophile. *Recherches sur les maladies chroniques et Analyse médicinale du sang*. Paris: Gabon, 1800 [1775].

de Fleury, Maurice. *Introduction à la médecine de l'esprit*. Paris: Félix Alcan, 1897.

Dehayes, P. M. "Chanson à digérer." *Le Gastronome*, no. 33 (11 July 1830).

Esquirol, Jean-Étienne. *Des passions considérées comme causes, symptômes et moyens curatifs de l'aliénation mentale*. Paris: Imprimerie Didot Jeune, 1805.

Esquirol, Jean-Étienne. *Des maladies mentales*. Vol. 1. Paris: J. B. Baillière, 1838.

Favre, Joseph. *Dictionnaire universel de cuisine et d'hygiène alimentaire. Modifications de l'homme par l'alimentation*. Paris: Les libraires, 1889.
Flaubert, Gustave. "Lettre à Edma Roger des Genettes." In *Correspondance*, edited by Jean Bruneau, vol. 4. Paris: Gallimard, 1998.
Flaubert, Gustave. n.d. "Dossier G226—Vol. 7." http://www.dossiers-flaubert.fr/cote-g226_7_f_173__r____. Accessed 28 June 2018.
Gauchet, Marcel, and Gladys Swain. *La Pratique de l'esprit humain*. Paris: Gallimard, 1980.
Goncourt, Edmond, and Jules de Goncourt. *Journal. Mémoires de la vie littéraire*. Vol. 2. Paris: Robert Laffont, 1989.
Goncourt, Edmond, and Jules de Goncourt. *Journal. Mémoires de la vie littéraire*. Vol. 3. Paris: Robert Laffont, 2004.
Guardia, Joseph-Michel. *La Médecine a travers les siècles: histoire-philosophie*. Paris: Baillière, 1865.
Guardia, Joseph-Michel. *Histoire de la médecine, d'Hippocrate à Broussais et ses successeurs*. Paris: O. Doin, 1884.
Huysmans, Joris Karl. *À rebours*. Paris: Gallimard, 1977 [1884].
Lombard, Léandre-Moïse. *Le Cuisinier et le médecin*. Paris: L. Curmer, 1855.
M. de V. "Les caprices de l'estomac." *La Gastronomie. Revue de l'art culinaire ancien et moderne*, no. 14 (5 January 1840): 2–3.
Marquer, Bertrand. "De l'épigastre au ventre: *œconomia animale* et économie du corps social." *Romantisme*, no. 154 (2011): 53–64.
Marquer, Bertrand. *L'Autre siècle de Messer Gaster. Physiologies de l'estomac dans la littérature du 19e siècle*. Paris: Hermann, 2017.
Monin, Ernest. *L'Hygiène de l'Estomac. Guide pratique de l'alimentation*. Paris: O. Doin, n.d.
Nordau, Max. *Dégénérescence*. 2 vols. Paris: Félix Alcan, 1894.
Nordau, Max. *Psycho-physiologie du génie et du talent*. Paris: Félix Alcan, 1897.
Raisson, Horace. *Code gourmand, Manuel complet de gastronomie, contenant les lois, règles, applications et exemples de l'art de bien vivre*. Bruxelles: Librairire Aug. Wahlen – Imprimerie de la Cour, 1828 [1827].
Réveillé-Parise, Joseph-Henri. *Physiologie et hygiène des hommes livrés aux travaux de l'esprit*. Paris: Baillière, 1881 [1834].
Seure, Jules. *Dyspepsie et dyspeptiques, étude pratique sur les maladies de l'estomac et des organes digestifs dans leurs rapports avec la dyspepsie, traitement alimentaire, médication proprement dite*. Paris: A. Coccoz, 1885.
Starobinski, Jean. "L'encre de la mélancolie." *La Nouvelle Revue française*, no. 123 (March 1963): 410–423.
Tampon, Colin. "Gendelettra morbus." *Le Figaro. Supplément littéraire du dimanche* (5 July 1890).

Tissot, André. *De la santé des gens de lettres*. Paris: Baillière, 1826 [1766].
Vigarello, Georges. *Les Métamorphoses du gras. Histoire de l'obésité*. Paris: Seuil, 2010.
Zola, Émile. *The Masterpiece*. Translated by Thomas Walton. Translation revised and introduced by Roger Pearson. Oxford: University Press, 1993 [1886].
Zola, Émile. *L'Assommoir*. Translated by Margaret Mauldon. Oxford: University Press, 2006 [1877].

CHAPTER 4

Situating the Anal Freud in Nineteenth-Century Imaginaries of Excrement and Colonial Primitivity

Alison M. Moore

Other chapters of this volume have shown that ideas about the gut flourished in multiple genres across continental Europe, Britain and the USA in the nineteenth century. But no thinker developed as much meaning in relation to the lower gut as the Viennese founder of psychoanalysis Sigmund Freud. The anal/excremental Freud can only be historically appreciated both by excavating his uptake of biological and ethnographic thought and by resituating him in the broad continental European *fin-de-siècle* milieu in which excrement had become the subject of an emergent field of new meanings. This was a field in which ideas about social progress, colonial power and class propriety were seen as given by a particular relationship of modern subjects to the lower gut. In this chapter, Freud's anal ideas are considered both in relation to ethnographic and biological texts that directly influenced his thought and in relation to cultural discourses and social pressures likely to have been at least

A. M. Moore (✉)
Western Sydney University, Penrith, NSW, Australia

M. Mathias and A. M. Moore (eds.), *Gut Feeling and Digestive Health in Nineteenth-Century Literature, History and Culture*, Palgrave Studies in Literature, Science and Medicine,
https://doi.org/10.1007/978-3-030-01857-3_4

partially responsible for his unusual theories of the role of the anus and excrement in both social evolution and individual psychic development.

One can trace a distinct genealogy of "psycho-anal" inquiry following from Freud's work, which would include the work of Sandor Ferenczi,[1] Karl Abraham[2] and particularly Julia Kristeva's 1980 *Pouvoirs de l'horreur* (*Powers of Horror*).[3] The most developed of these engagements was Kristeva's; however, her discussion of abjection appears to have been based on her uptake of multiple other philosophical and anthropological approaches such as those of Georges Bataille, Henri Lévi-Strauss, René Girard and Mary Douglas as much as with Freud's anal speculations. In fact, her reliance on anthropological thought to theorise about abjection in Freudian terms itself provides a clue to how Freud may have developed these ideas in the first place. This is not because defecatory practices were central to the history of ethnographic and anthropological observation per se. In fact, as the Dutch medical anthropologist Sjaak Van der Geest has remarked, it is probably more curious that so few mentions were ever made of excretory customs of the peoples studied by European anthropologists throughout the nineteenth and twentieth centuries, given the striking variation across cultures in norms and rituals relating to such bodily functions.[4] But confronted with excremental culinary, ritual and healing practices in diverse cultures around the world, some nineteenth-century ethnographic writers clearly converged on a model of how the treatment of excrement could be taken as sign of social development, imagined according to a uni-linear teleology from primitive, to barbarous to civilised stages. These ideas impacted Freud in his peculiar fusion of evolutionary biology with a view of psychosexual individual development. But broader mentalities circulated in Freud's youth that also related the modern struggle with excrement to the path of civilisational progress, including discourses of progress generated in the sanitation crises and sewerage reconstruction projects of major European cities, as well as hygienic ideas about constipation in modern life.

[1] Ferenczi, "The Ontogenesis of the Interest in Money" (1916).

[2] Abraham, "A Short Study of the Development of the Libido" (1926).

[3] Kristeva, *Pouvoirs de l'horreur*.

[4] Van der Geest, "Not Knowing about Defecation," 75–76. See also Moore, "Historicizing the Modern Excremental Worldview".

The approach in this chapter reflects a variety of methodologies derived from intellectual history in the Cambridge-style tradition,[5] cultural history of *mentalités* in the *Annales* tradition,[6] as well as the excavation of temporal layers in the history of concepts in the *Begriffsgeschichte* tradition.[7] This methodological fusion is necessary because Freud's ideas about the anus had partially distinct textual origins, but were also stimulated by ambient mentalities at the broad cultural level, and involved a unique crystallisation of these two different kinds of forces into a novel coherent concept. The sources on which I focus here are primarily psychoanalytic, medical and ethnographic. Literary sources certainly were a part of this broad cultural ambiance, and although it is beyond the scope of this chapter to discuss them, many of the authors who treated matters of filth and sewers, particularly Victor Hugo and Émile Zola, were well known to Freud also.[8]

Freud was a highly cosmopolitan thinker who read works in German, French and English, and took casual interest in a wide variety of knowledge that included biological evolutionary thought, ancient history, mythology, ethnography, literature, as well as the neurology and psychiatry that formed the basis of his own medical training. And it was precisely this breadth of perspective that gave psychoanalysis such a powerful appeal across multiple other disciplines, many of them founded in the humanities.[9] Germanic folk culture has a long history of scatological humour, as Alan Dundes has shown,[10] and Freud may well have been exposed to such humour in his youth. But his ideas about the anus went far beyond the ribald content of this style of expression. His early medical education in the 1870s and 1880s was conducted at the University of Vienna, and his professional induction as a neurologist occurred in Paris at the *Salpêtrière* hospital during the late 1880s, under the renowned French "hysteria" specialist, Jean-Martin Charcot.[11] Much of his final decade of life (from 1929 to 1939), during which his ideas

[5] Whatmore, *What Is Intellectual History?* 45–55.

[6] Burke, *What Is Cultural History?* 4, 33.

[7] Koselleck, *The Practice of Conceptual History*, 138–139.

[8] See Gay, *The Cultivation of Hatred*, 178–180; Gay, *Reading Freud*, 107–110.

[9] See Sulloway, Freud, *Biologist of the Mind*, xii.

[10] Dundes, *Life Is Like a Chicken Coop Ladder*.

[11] Gay, *Freud, A Life for Our Time*, 44–62.

about civilisation and the body were crystallised, was spent in London. Situating this aspect of his thought therefore requires traversing several cultural, linguistic and disciplinary domains.

I

As Ranjana Khanna and Jean Walton (as well as some of my own previous work) have shown, Freud's highly gendered account of male psychosexual development was imagined mirroring the historic process of colonial exploitation.[12] But he indicated in texts beginning in the early 1900s that it was the pleasure of defecating and of anal sensation generally that the civilised individual first needed to sublimate in order to form both a moral conscience and a sense of shame that would ensure social conformity to middle-class life. It was the symbolic basis of induction into a world of monetary value, producing the desire to engage, as an adult, in relations of finance and commodity exchange. Without most of the men in a culture transacting towards their bodies in this fashion, civilising progress would not be possible. "Anal erotism" he claimed, "succumbs in the first instance to the 'organic repression' which paved the way to civilisation".[13]

He appears to have arrived at the idea of money/modernity as the sublimation of excrement early in his clinical practice and wrote to his mentor Wilhelm Fliess in 1897 discussing the German linguistic slang referring to someone who hated to spend money as *filzig* (just as we speak of the "filthy-rich" in English) and to a wealthy spendthrift as a *Dukatenscheisser* (shitter of ducats).[14] One year before this, he had also suggested in a letter to Fliess, the notion of the "erotogenic zones", of which the anal was the most primitive.[15] Around the same time, he had asked Fliess "in connection with the eating of excrement" if there was ever a phase in a child's development when disgust in such things was not yet developed, adding that "the answer would be of theoretical interest".[16] He clearly found his answer to this question, as indicated in later

[12]Khanna, *Dark Continents*, 48–52; Walton, *Fair Sex, Savage Dreams*, 70–74; and Moore, *Sexual Myths of Modernity*, 55–80.

[13]Freud, *Civilisation and Its Discontents*, 47 ftn.

[14]Freud, "Character and Anal Erotism," 295–297.

[15]Strachey, "Editor's Note," 5.

[16]Letter of February 8, 1897. *The Cambridge Companion to Freud*, 182.

statements he made that "the excreta arouse no disgust in children... and seem valuable to them as being part of their own body which has come away from it".[17] It therefore came as no surprise to him to discover excremental practices among "primitive" cultures found in ethnographic descriptions circulating around this time—the European child was passing through the primitive evolutionary phase on the path to civilised adulthood.

In the 1908 article "Character and Anal Erotism", Freud discussed his attempts at treating patients whom he described as "especially orderly, parsimonious and obstinate" (who he designated as the "anal" character type). Such individuals, he claimed, confessed to having been highly resistant to toilet training as children and had been inclined to do "all sorts of unseemly things with the faeces" once passed. These were people for whom the "primitive" pleasure of retention and defecation was particularly strong, and so required as adults that their anal pleasure be sublimated into a neurotic form of behaviour, resulting in excessive concern with cleanliness and control.[18] But the sublimation of anality was essential even in the normal masculine Œdipal process. Defecation was a primary stimulus in the development of the psyche, since the Œdipus complex began when the child was first required to defecate on a toilet or chamber pot instead of in nappies.[19] The child sees that a phallic-shaped object (his turd) comes away from his body, is detached, removed and lost to him. From this comes his first suggestion of the threat of castration, which becomes definitively reinforced by his realisation that girls are indeed already castrated. The girl child must transform both her recognition of castration and the sense of loss of the turd ("*lumph*") into the maternal desire for a baby. The boy child's development of a functional sense of shame and conformity depended upon the recognition that his faeces, as well as the sexual desire for the boy's mother, must be given up as a trade for maintaining the pleasure of the phallus.[20] Hence, the faeces that one "gives up" to the potty must be also metaphorically rescinded and sublimated into a new form of desire. Since the child's faeces were its first "gift" to the mother, it must later

[17] Freud, *Civilisation and Its Discontents*, 47 fnt.

[18] Freud, "Character and Anal Erotism" (1908), 294–297.

[19] Freud, "The Dissolution of the Oedipus Complex" (1925), 663. See also Freud, "On the Transformation of Instincts with special reference to Anal Erotism" (1917), 171.

[20] Freud, "The Dissolution of the Oedipus Complex," 661–664.

transmute anal pleasure into a pleasure in gifts, which later becomes an interest in money (for boys) or in child rearing (for girls).[21]

Freud argued that there were essentially two challenging experiences faced by all children in civilised life: "the withdrawal of the mother's breast...and the daily demand on them to give up the contents of the bowel".[22] Defecation was thus the primary ground upon which a child learnt to choose object-focused love from a narcissistic anal-masturbatory love:

> He either parts obediently with his feces, 'offers them up' to his love, or else retains them for purposes of auto-erotic gratification and later as a means of asserting his will. The latter choice constitutes the development of defiance (obstinacy), a quality, which springs, therefore, from a narcissistic clinging to the pleasure of anal erotism.[23]

In a normative development, anal erotic urges would be sublimated to produce a functional individual within the bourgeois economic order. Modern life was dependent upon the Œdipal rescinding of anal pleasure. Constipation, Freud noted, could be treated by looking at a patient's relationship to money. In ancient Babylonian mythology, he claimed, gold was called *ilu manman*, or "feces of Hell". He added: "It is possible that the contrast between the most precious substance known to men and the most worthless, which they reject as waste matter ('refuse'), has led to this specific identification of gold with feces".[24] In later editions of *The Interpretation of Dreams*, he claimed that dreams about defecating, both his own and his patients, tended to illustrate this symbolic connection with money "which is also supported by copious evidence from social anthropology".[25]

Freud's ideas about sublimation of anal sensation remained part of his account of the development of civilisation and of the individual through to the end of his life. In *Civilization and Its Discontents* (1929), he referred still to the idea that "order and cleanliness are important

[21] Freud, "On the Transformation of Instincts with Special Reference to Anal Erotism," 168.

[22] Freud, "The Dissolution of the Oedipus Complex," 175.

[23] Freud, "On the Transformation of Instincts," 168.

[24] Ibid., 297.

[25] Freud, *The Interpretation of Dreams*, 413, 475.

requirements of civilization". Sublimation of anal instinct, which he likened to the "process of civilization" itself, was what made "higher psychical activities, scientific, artistic or ideological" possible.[26]

II

While Freud's ideas about the role of the anus and excretory function in psychic development displayed a peculiar application of evolutionary thinking to a theory of the psychosexual body, he was certainly not alone in viewing excretory practices as inscribed within the civilised stage of social development imagined according to a social evolutionary teleology—as a process of self-mastery akin to mastery over others in colonial conquest. Freud's own youth and training as a young neurologist was spent in major European cities (Vienna and Paris) where common cultural concerns about sewerage, foul odours and middle-class propriety indicated a broad conceptual layer of meaning connecting the conquest of excrement with the triumph of European modernity. These ideas had formed in a succession of sewerage crises beginning early in the nineteenth century but recurring also in the late nineteenth century as new waves of urban population growth pressured existing city infrastructures yet again.

In 1838, the urban engineer Victor Considérant referred to Paris as itself a living body that must breathe, but was suffocating from the "immense studio of putrefaction" that the city's cesspools had created.[27] The mid-century technologisation of the Paris sewerage system under the Second Empire of Louis Napoléon III tackled rising disease epidemics that were the product of further urban intensification. Town planners and engineers discussed the crisis in terms that tended to relate cleanliness to the masculine conquest of excrement as a sign of the civilising process.[28] A *Bulletin de l'Industrie de Paris* published in 1863 urged the municipal government, with its council "of the most capable men of science", to live up to the

[26]Freud, *Civilisation and Its Discontents*, 44.

[27]Considérant, *Description du phalanstère et considérations sociales sur l'architechtonique*, cited in Guerrand, *Les Lieux*, 77.

[28]Corbin, *Le Miasme et la jonquille*, 167–188; Delaporte, *Le Savoir de la maladie*, Ch. 5; also Laporte, *l'Histoire de la merde*; Reid, *Paris Sewers and Sewermen*; Barnes, *The Great Stink of Paris*.

image of Paris as representing "progress in all things".[29] The Paris Inspector of Public Works at this time was ironically named Pierre-Émmanuel Bruneseau (which means 'brown-waters'!). Victor Hugo's 1862 novel *Les Misérables* referred to him as the "Christopher Columbus" of the "cloaca".[30] Donald Reid's history of the Paris sewers remarks on a distinct mentality that emerged in the time of the *Préfet de la Seine* Baron Georges Haussmann, in the reconstruction of Parisian streets and underground spaces. The word "*cloaque*" (cloaca, the singular excretory, urinary and generative orifice of birds) with its biological connotation was increasingly replaced by the term "*égout*" (sewer), connoting a technological triumph.[31] Commonly when French writers discussed the virtues of Parisian sewer technology, it was in a tone of national pride. But such hubris did not necessarily reflect the resolution of the city's recurring problems of sanitation.[32] Alfred Mayer in his entry in the 1867 *Guide-Paris*, even before the completion of the major reconstruction of the Haussmann years, had described the sewers of Paris at that time as indicating that French civilisation had at last surpassed even the grandeur of Ancient Rome—that other great civilisation known for its sewers and aqueducts.[33]

Similar associations were often made in England between civilisation and mastery of excremental pollution. Henry Mayhew's sociological study of the London poor in the 1850s emphasised the "filth" entailed in urban poverty, focusing on "nightmen" and other cesspool workers.[34] In the context of these genuine challenges to urban infrastructure in nineteenth-century cities, excremental odours were a recurrent feature of city life, making the symbolic universe of excrement particularly vivid to thinkers of this time. Christopher Herbert notes the widespread associations of money with excrement in a range of Victorian thinkers from Dickens (in his depiction of the "filthy" rich Mr Merdle in *Little Dorrit*) to Henry Mayhew (in his equation of the wealth of Britain with

[29] Anon, *Bulletin de l'Industrie de Paris*, cited in Anon, *Seule réponse à l'article anonyme*, 5.

[30] Hugo, *Les Miserables*, 879.

[31] Reid, *Paris Sewers and Sewermen*, 36.

[32] Barnes, *The Great Stink of Paris*.

[33] Mayer, "The Canalisation souterraine de Paris," 184.

[34] Mayhew, *London Labour and the London Poor*, vol. 2, 386.

the abjection and poverty of its lower classes).[35] Gertrud Himmelfarb has noted the recurrent metaphor of money as filth in the opus of Karl Marx too.[36] Throughout the nineteenth century, urban population intensification and industrialisation meant that the larger European cities were repeatedly confronted with excrement as an unprecedented social problem.[37] A preoccupation of some kind then was perhaps inevitable, but practical challenges alone cannot explain the specific meaning content of these concepts.

By the end of the nineteenth century, the Paris sewers were transformed into such a mark of civilised pride that they could be opened for public tourist exhibition in the first of what became an enduring tradition of the "*visite à l'égout*".[38] The late nineteenth-century cultural historian Alfred Franklin devoted an entire volume to the history of hygiene in his 7-volume 1890 study of *La Vie privée d'autrefois* (*Private Life in Olden Times*), showing how the growth of glory of the French state and the moral rectitude of the French people coincided with the banning of public defecation and urination, with the deodorisation of Paris and with the creation of an efficient sewage system. In Franklin's vision, the old dirty, unrestrained habits of the medieval past were clearly bound up with an imagined lack of "civilisation" understood in both socio-economic and moral terms.[39] Franklin claimed that the first toilets in France were constructed in churches under the reign of Louis XVI and were referred to as "des lieux à l'angloise", or "English-style places".[40] He recounted the tale of a doctor Rigord who served the monarch Philippe-Auguste, and who in 1185 claimed that the very first construction of public works by the French monarchy originated from a single incident in which the dauphin had choked with disgust from the odour of the town muck in which the wheels of his carriage had become stuck, provoking his ordonnance to pave the streets of Paris immediately.[41] In Franklin's view then, disgust at excrement was the driver of modernising development.

[35] Herbert, "Filthy Lucre," 185–213.

[36] Himmelfarb, "The "Real" Marx," 39.

[37] Corbin, *Le Miasme et la jonquille*; Barnes, *The Great Stink of Paris*.

[38] Reid, Paris Sewers and Sewermen, 39–50.

[39] Franklin, *La Vie privée d'autrefois*, Tome VII.

[40] Ibid., 176.

[41] Ibid., 5–6.

Late nineteenth-century Europe also saw the emergence of an unprecedented fixation with lower intestinal health. Stimulant sweets, coffee and tea, new consumer commodities all derived from colonial exploitation of South American, African and Indian produce, were increasingly used by Europeans as valuable props generally for stamina in the working day, or more particularly as cures for constipation or urinary stagnation.[42] Hygiene manuals and medical writings emphasised intestinal stasis as the source of all ill health, advising readers in the use of purgative medicines, enemas, diet, exercise and lifestyle control, often depicting constipation as the quintessential disease of the civilised European due to the sedentary lifestyles of the burgeoning middle classes.[43] As medical historian James C. Whorton remarks, we should not assume the depth of concern expressed about constipation in the nineteenth century to be exaggerated.[44] Such preoccupations with constipation and intestinal health may well have owed part of their intensity to the very real changes in work practices and lifestyles with the growing mechanisation of labour and the burgeoning of middle-class work patterns, producing more sedentary bodily practices that promoted intestinal stasis.[45]

This, then, was the broad cultural context in which Freud's ideas about anal sublimation and constipation emerged. In reflecting on one of his own excremental dreams in the eighth edition of *The Interpretation of Dreams*, Freud remarked that he had been stimulated in its content by perusing, before bed that night, Jules Garnier's catalogue of the works of Rabelais (illustrated by Gustave Doré) in which Gargantua is pictured urinating over Paris from his seat atop the Notre Dame cathedral [see image].[46] Freud had dreamt that he was washing away a pile of excreta with a stream of his own urine, experiencing a feeling of triumph and mastery in doing so. On awakening, he recalled that the area around Notre Dame had been one of his favourite haunts while studying in

[42] Fauvel, "Action du chocolat et du café sur l'excrétion urique," 854. Also: Choqart, *Aux consommateurs de chocolat et de thé*, 39.

[43] For a detailed discussion of this phenomenon see Whorton, *Inner Hygiene*; also Frexinos, *Voyage sans transit*.

[44] Whorton, *Inner Hygiene*, xii.

[45] Ibid., 33.

[46] Garnier, *L'Œuvre de Rabelais* (Paris: Block, 1889).

Fig. 4.1 Gustave Dore's illustration of Gargantua for the 1928 Garnier edition of Rabelais' text that Freud read before the dream described in *The Interpretation of Dreams*, 8th edition. Wikimedia Commons

Paris with Charcot. "The fact that all the faeces disappeared so quickly under the stream" of the sewer made him recall the biblical expression "*Flavit et dissipati sunt*" [he blew and they were scattered].[47] Freud's sense of triumph reflected precisely the wishes of Parisian engineers to conquer the filth of that city, to colonise the malodorous remnants of the pre-modern metropolis (Fig. 4.1).

[47] Freud, *The Interpretation of Dreams*, 475–476. This work was first published in 1899. The fourth edition cited here was published in 1928.

III

From the perspective of intellectual genealogy, it seems likely that Freud's ideas about defecation came most directly from two specific branches of thought. The first of these was a form of biological evolutionary speculation found in popular accounts of embryonic phylogeny. The second was late nineteenth-century ethnographic description of bodily practices in hunter-gatherer societies. Freud's vision of psychic development as echoing the evolution of culture from primitive to civilised was modelled on Ernst Haeckel's widely influential account of biological phylogeny in the 1870s. Freud referred explicitly to Haeckel in a 1913 work,[48] but as Patricia Cotti notes, he had discussed Haeckel's specific ideas without attribution as early as 1908.[49] Haeckel was responsible for the term "gastrulation" in descriptions of early animal embryonic development and the formation of the mouth–anus "blastopore", or precursor of the gut (terms which have since persisted in the vocabulary of modern embryonic physiology).[50] In this model, the gut is the first feature of animal physiology to evolve, and the most primitive filter-feeding animals are those with little else than a mouth, anus and a digestive tube that connects the two orifices. The mouth, anus and gut were those most primitive features of human anatomy, the most conserved from the evolution of the first animal life on earth. But importantly for Freud, Haeckel also advanced the notion of genetic recapitulation, summarised in the expression "ontogeny recapitulates phylogeny", or the idea that each organism in embryonic development passes through the earlier stages of its own prehistoric biological evolution.[51] Freud's innovation was to apply that vision of biological evolution to a presumed social evolution of human experience of the anus, which was then recapitulated in each individual's psychic development.

As Frank Sulloway notes, Freud's engagement with Haeckel was probably mediated, at least in part, by the German popular science vulgariser and biographer of Haeckel, Wilhelm Bölsche. Bölsche's grand multivolume *Das Liebesleben in der Natur* (1898–1903) was a work which Freud

[48] Freud, "The Claims of Psychoanalysis to Scientific Interest" (1913), 168.

[49] Cotti, "Sexuality and Psychoanalytic Aggrandisement," 64.

[50] Glick, *The Comparative Reception of Darwinism*, 110. Haeckel, *The Riddle of the Universe*.

[51] Gould, *Ontogeny and Phylogeny*, 1.

held in his London library, and cited in his Introductory Lectures.[52] In this work, Bölsche particularly lingered over Haeckel's description of the mouth–anus blastopore in the development of the animal zygote into a multicellular organism.[53] Bölsche indeed elaborated what was merely implicit in Haeckel—that anal, oral and sexual development was the primary field of growth of the emergent organism. This suggested to Freud the necessary primacy of all sensations associated with the anus, the mouth and the genitals in early human experience.[54] Childhood was primitivity, in which the ancient stages of animal and human evolution lingered most clearly.

This idea that each individual experienced in psychic development all the stages of ancient biological and social evolution formed a complex network of conceptual layers that Freud reinforced with reference to ethnographic descriptions. Childhood development and social evolution were seen as directly comparable because each psyche had to recapitulate as early animal, then primitive human, and finally civilised individual in a succession of stages. Assuming social development to follow a uni-linear course was something that Freud's contemporary, the great anthropologist Franz Boas in 1909 had specifically rejected, arguing that each culture lives its own unique development that cannot be assimilated to European historic patterns.[55] Freud would certainly have known of Boas' ideas, but appears to have found the uni-linear model, as assumed in his 1913 work *Totem und Tabu*, more generative for his own speculations about phylogenetic recapitulation.[56] The idea that appears to have particularly confirmed Freud's views about primitivity and anality was the notion found in a number of ethnographic and anthropological works published at the turn of the century, that primitive societies all shared a lack of taboo surrounding excrement. Did children not also require discipline to manage their defecatory waste?

As the feminist race scholar Anne McClintock notes, in nineteenth-century ideas "...evolutionary progress was represented as a series of anatomically distinct family types, organised into a linear procession,

[52] Sulloway, *Freud, Biologist of the Mind*, 261–262; Gould, *Ontogeny and Phylogeny*, 157.

[53] Bölsche, *Das Liebesleben in der Natur*. See Sulloway, *Freud, Biologist of the Mind*, 264. See also Amouroux, "Le Précieux livre de W. Bölsche," 24–36.

[54] Sulloway, *Freud, Biologist of the Mind*, 262–263.

[55] See Kenny, "Freud, Jung and Boas," 4; Boas, *The Mind of Primitive Man*.

[56] Discussed in Moore, *Sexual Myths of Modernity*, 67–68.

from the 'childhood' of 'primitive' races to the enlightened 'adulthood' of European imperial nationalism".[57] Late nineteenth-century ethnographers studying Australian Aborigines conveyed this prevalent view of colonially subjected peoples as the naïve children of imperial Europe. The childishness of "primitive" humans was often imagined idealistically, even sympathetically, as a form of innocence from modern vices and perversions. The French ethnographer and anarchist Élie Reclus was one who considered the destruction of Australian Aboriginal cultures in the British colonial conquest to be a true loss for humanity, seeing them as essentially infantile and innocent. Reclus' 1894 work *The Australian Primitive* (*Le Primitif de'Australie*) included chapter headings referring to "simple souls", and the "negroe's naivety".[58] In associating Aborigines with childlike innocence he also imagined their "primitivity" as entailing a greater engagement with the body, both its pleasures and its products. Reclus detailed practices among Ngarrindjeri ("Narrinyeri") people of South Australia in which excrement was used in ritualistic contexts, describing mourning women at the death of a family member, having already covered themselves in "muck" (fange), "they cover their heads with excrement" such that "their tears carve furrows through their mourning mask".[59]

But Reclus also reflected more broadly on the concept of "Divine excrement" in various pre-modern and extant hunter-gatherer cultures, as if both were examples of some similar lack of "civilisation". He described initiation rituals in various Australian and African hunter-gatherer cultures, in which the initiate would be beaten, defiled, and only considered to have passed the ordeal once becoming thoroughly malodorous. "Let us recall though that in several of the ancient guilds of Germany and of Hungary, the apprentice only became peer via a malodorous initiation", he added, then citing an expression supposedly current among Sicilian fortune-tellers: "Whoever eats the Gypsy's shit becomes divine".[60] He described the Japanese myth of Iza Nami

[57] McClintock, *Imperial Leather*, 359.

[58] Reclus, *Le primitif d'Australie*, 131, 339. "[des] âmes simples", "[la] naïveté du nègre".

[59] Ibid., 227–228. "elles se couvrent la tête d'excréments...leurs larmes creusent des sillons dans le masque de deuil".

[60] Ibid., 199. "*Cui mancia mmerda di Zingari diventa indivinnu*"; "Rappelons cependant qu'en plusieurs anciennes corporations de l'Allemagne et de la Hongrie, l'apprenti ne passait compagnon que par une initiation mal odorante".

who was thought to have excreted both the earth in its solid form and the ocean as liquid; adding that "in Polynesia, the tombs are consecrated with this substance", and that in southern Africa, certain hard nuts are only eaten after being pre-digested by a bird.[61] The Ygarrotes of the Philippines poured "the juice of the innards" of their fresh kill over the meat as a sauce.[62] "The notion of 'divine excrement', moreover used by innocent pantheists of the black race" implied the idea of "celestial Emanation" which, he remarked European theologians had borrowed from Eastern philosophies and religions.[63] The use of excrement for sacred ritual then was a primitive inheritance that Christianity had domesticated as the Holy Communion.

The Scottish Australian grazier and amateur ethnographer Peter Beveridge, studying the Australian Aborigines of the Victorian Riverina area in the 1880s, also explicitly linked infantile "primitivity" to the cultural use of excrement for medicinal purposes. His account of the culture of the Riverina tribes was published on the request of his widow following his death in 1885.[64] He described Riverina Aborigines as under threat of extinction, in part because of their own "entire lack of restraint, when any one of their animal instincts chances to be in the ascendant".[65] This applied both in their enthusiasm for the "fire-water" that Europeans brought to the continent, as much as in the satisfaction of all corporeal desires, such as those for food or for sex.[66] "When food is plentiful they feast and riot to the top of their savage heart, gorging themselves...until their abdominal regions become so distended as to be decidedly uncomfortable". Like the wicked grasshoppers of Aesop's tale, "they never try to preserve any for future use, but allow everything which they cannot stuff into themselves to go to waste". In fact, so used to the smell of rotting food was the "blackfellow" that "his olfactories are none of the most sensitive".[67] "Chastity is quite unknown to them", he noted.[68]

[61] Ibid., 199.

[62] Ibid., 31.

[63] Ibid., 200.

[64] Beveridge, *The Aborigines of Victoria and the Riverina*, 4.

[65] Ibid., 6.

[66] Ibid., 7.

[67] Ibid., 9–10.

[68] Ibid., 23.

And since "intercourse is quite unrestricted between the sexes, it can very easily be imagined how this foul malady runs riot…"—this, he thought, was one of the causes of their relative sterility and degeneration.[69] He then described (in Latin, on account of its shocking quality) a "most disgusting remedy" he reported witnessing, involving the use of communally collected excrement ingested as a cure for a sick and dying person. "The aborigines have an unbounded faith in this truly horrible dose, and enumerate many, many instances wherein it has effected marvellous cures"; although, he added, he knew of cases where it had not worked at all. He noted that "some *savants* positively assert" that excrement was "the very essence of life, as well as containing the germs thereof", and that is was thus conceivable it may act as cure for someone on the brink of death; "but this is purely for physicians to consider", he concluded dismissively.[70] In fact, the use of excrement in medicinal remedies was also a common practice described in early modern pharmacological texts from across Europe (possibly what Beveridge was alluding to in his mention of the "savants").[71]

From the perspective of our current medical models, such practices across cultures may indeed be understood as medically intuitive and by no means indicative of cultural simplicity or infantilism. But such interpretations were not available to ethnographers in the late nineteenth century. Gastroenterology researchers since the late twentieth century have increasingly emphasised diversity of the gut microbiota to be a crucial marker of health, with research on gut conditions such as Crohn's disease and ulcerative colitis correlated with the loss of microbial diversity.[72] Since the end of the twentieth century, faecal transplants that inoculate the gut with more diverse microbes have been used therapeutically in the treatment of such diseases with a success-rate far exceeding any pharmacological agents, with delivery methods varying from rectal to oral entry depending on the part of the intestinal tract where microbial imbalance

[69] Ibid., 12.

[70] Ibid., 53.

[71] LaGuardia, "Doctor Rabelais and the Medicine of Scatology," 38–54; Schmidt and Simon, "Holy and Unholy Shit," 109–119.

[72] Lloyd-Price, Abu-Ali and Huttenhower, "The Healthy Human Microbiome," 1–11; Guinane and Cotter, "The Role of the Gut Microbiota in Health and Chronic Gastrointestinal Disease," 296.

has been identified.[73] Faecal transplant therapy is one of the few effective treatments for the dangerous bacterial infection with *Clostridium difficile*.[74] And as antibiotic-resistant strains of bacterium have now become increasingly powerful, common and worrisome, non-antibiotic therapies reflecting a holistic view of human-bacterial interactions, are attracting new clinical interest.[75] Ethno-microbiologists such as Jeffrey Leach have recently shown that the West African Hadza tribespeople's ingestion of faecal material from the intestines of slaughtered deer both contributes to their gut microbial diversity, and allows them to access the cellulose-rich green plant matter fermented by deer's ruminant digestive tract, making it bioavailable for human consumption, which it would not otherwise be as grass. This is thought to be responsible for the generally good health and longevity of this subsistence-living population in a harsh environment where green vegetable matter digestible directly by humans is absent.[76]

But the Hadza people now studied by twenty-first-century microbiologists as models of gut health precisely because of their microbial diversity and environmental adaptability were also one of the examples cited in a late nineteenth-century work that speculated about the excremental practices of "primitive" peoples—a work in which Freud took great interest. US cavalryman Captain John G. Bourke spent the eighteen-eighties studying the excremental practices of various "non-civilised" peoples. His 1891 *Scatalogic Rites of All Nations* was the result.[77] This 450-page text exhaustively surveyed ethnographic and anthropological descriptions of coprophagic rituals throughout the world. Peter Beveridge, and especially Élie Reclus, were among the authors he cited

[73] Lopez and Grinspan, "Fecal Microbiota Transplantation for Inflammatory Bowel Disease," 374; Ho Bak, et al., "Fecal Microbiota Transplantation for Refractory Crohn's Disease," 244.

[74] Lehrer, "Duodenal infusion of feces for recurrent Clostridium difficile," 2144.

[75] Hecke, Wang, and Lee et al., "Implications for Antibiotic Resistance for Patients' Recovery from Common Infections," 371; Bernabé, Langendorf and Ford et al., "Antimicrobial Resistance in West Africa," 30274; Le Page, Gunnarsson and Snape et al., "Integrating Human and Environmental Health in Antibiotic Risk Assessment," 30900.

[76] Leech, "Gut Microbiota," 1; Schorr et al., "Gut Microbiome of the Hadza Hunter-Gatherers," 3654.

[77] Bourke, *Scatalogic Rites of All Nations.*

repeatedly throughout his book.[78] The subject, he remarked in the preface, "however repellent it may be", warranted

> the profoundest consideration, if for no other reason than that from the former universal dissemination of such aberrations of the intellect, as well as of the religious impulses of the human race, and their present curtailment and restriction, the progress of humanity upward and onward may best be measured.[79]

Excremental practices then formed part of universal primitive impulses, and the sacralisation of excrement in religious ritual signalled the beginning of the march towards civilisation. Finally, as the reign of religion gave ground to modern civilisation and science, such practices lost their sacred allure. Bourke read numerous second-hand accounts of the descriptions of sixteenth-century Spanish explorers, who depicted native American cultures of Lower California as eating all manner of vile things including spiders, beetles, earth, wood and "the dung of wild beasts".[80] Casteñeda de Nagera described the natives of Arizona in 1540 as "a race of naked savages who ate their own excrement".[81] It certainly did not appear to have occurred to Bourke that some of those accounts may have been exaggerated in a culture of justification for the colonial conquest. "Primitive" people, in Bourke's view of progress, could be expected to revel in excrement, such as the South and North American hunter-gatherers described by the colonial explorers.

But it troubled him that some of the great civilisations nonetheless also had their own examples of excremental remedies, and he concluded that this "persistence with which the same remedies have been perpetuated through centuries" derived from European Christendom's inheritance of a belief that all diseases were "punishments inflicted by angry gods", and so their remedies must take the form of "oblations or sacrifices to the propitiate the offended spirits".[82] In primitive medicine everything was included in the "*Materia Medica*": "Saliva, the ordure,

[78] Ibid., on Reclus: 36, 145, 170, 178, 189, 204, 209, 236, 271, 339, 348, 351, 392, 400, 430, 440, 446, 463; on Beveridge: 35, 355, 465.

[79] Ibid., v.

[80] Ibid., 33.

[81] Ibid., 34.

[82] Ibid., 456–457.

urine, catamenial fluid, blood, bile, calculi, bones, skulls – all were mysterious, and therefore were 'medicine', especially when obtained from a saint or lama".[83] Similarly, the medical practice of examining the urine and faeces of the sick "in earlier stages of human progress...was complicated with ideas of divination and forecast".[84]

In detailing the prevalence of excremental practices across human diversity, Bourke claimed to be continuing an important intellectual tradition, citing the early eighteenth-century natural philosopher Robert Boyle, who had indeed noted sceptically that "a despised common sample, nay an infect or an excrement may in some cases prove nobler medicines than an extract, elixir, or a quintessence".[85] (Though, it was important, Boyle had noted elsewhere, not to eat the excrements of the mad, lest one become mad oneself.[86]) The intellectual genealogy, by Bourke's account also included Martin Schurig's *Chylologia Historico-Medica* (1725), Michael Etmuller's *Opera Omnia* (1690), Franz Christian Paullini's *Dreck Apotheke* (*Filth Pharmacy*, 1696), and various ancient authorities who had deemed the topic of excrement worthy of medical consideration, in addition to the self-taught nineteenth-century historian Henry Thomas Buckle, who in 1872 had sketched seventeenth and eighteenth-century English moral attitudes to farting.[87] Curiously though here Bourke conflated those who described the history of scatologocial matters (e.g., Buckle and himself) with those who prescribed them medically: Both Schurig's *Chylologia* and Paullini's *Dreck Apotheke* were precisely such works detailing how to prepare excremental remedies and for which conditions they should be ingested.

Many of Bourke's examples were drawn from an earlier French work of 1849 entitled *Bibliotheca Scatologica*, by Auguste Veinant, Pierre Jannet and Jean-François Payen.[88] (The latter was a doctor in the Haute Savoie region who treated his patients with prescriptions for thermal baths, and collated the manuscripts and correspondence of

[83] Ibid., 400.

[84] Ibid., 272.

[85] Boyle, *The Usefulness of Experimental Philosophy*, 64.

[86] Boyle, *The Nature, Properties, and Effects of Effluvia*, 436.

[87] Bourke, *Scatalogic Rites of All Nations*, v; Buckle, *Miscellaneous and Posthumous Works*, vol. II, 472.

[88] Veinant, Jannet and Payen, *Bibliotheca Scatologica*; citations in Bourke, *Scatalogic Rites of All Nations*, iv, 30–32, 46, 114, 120, 129, 134–137, 140, 156, 278–279, 341.

the sixteenth-century French philosopher Montaigne.[89]) This bibliography too was a sort of excremental compendium, which included short scatological verses in French, German and Latin, and an international bibliographic listing of poetic, literary, ethnographic and medical works published on the topic of defecation. It was framed as a necessary humanist corrective to the oversight in knowledge derived from religious presumptions that God had created nothing dirty.[90] "All that touches Man is worthy of study", the authors declared, since "nothing is more human than the *caput mortem* [worthless remains] of intestinal chemistry".[91] The authors noted the distinct European historical tradition of using excrement in medications, citing Schurig's *Chylologia* at length, and adding "The usage of *stercora* [manure], particularly that of men, as pharmaceuticals, is very real. We name those who prescribe them *stercoral doctors*".[92] They claimed that once they began to seek-out works to fill the bibliography, they found the "richness of their harvest" exceeded all of their hopes.[93] In classifying these scatological works, they imagined their bibliography to ennoble "these bastards of literature" and "civilise them".[94] Similarly, Bourke remarked on the ubiquity and plethora of examples available across cultures of using excrement for ritual, food or medical purposes—an important indicator of it forming a universal stage in the development of human progress.[95] He credited the *Bibliotheca Scatalogica* as being a "very learned pamphlet" and "curious book, full of learning and research", in spite of its occasional descents into ribald humour.[96]

But between the 1849 publication of the *Bibliotheca Scatologica* and the 1891 publication of Bourke's *Compilation of Notes and Memoranda*, an important conceptual change had occurred. No longer were excremental texts to be read as tokens of a materialist inclusivity positioned against religious refusal of the body. Instead, in Bourke we see the

[89] Payen, *Essai sur les eaux minérales*; Payen, *Notice bibliographique sur Montaigne.*

[90] Veinant, Jannet and Payen, *Bibliotheca Scatologica*, xv.

[91] Ibid., xix. "tout ce qui touche à l'homme est digne d'étude….rien n'est plus humain que le *caput mortuum* de la chimie intestinale…."

[92] Ibid., 29.

[93] Ibid., xx.

[94] Ibid., xv.

[95] Bourke, *Scatalogic Rites of All Nations*, iv.

[96] Ibid., iii, 31.

notion that organised religions, developing in "communities in a low state of culture", exhibited irrational beliefs about excrement as a potentially holy substance which were carried over into early modern medical traditions.[97] The ordure of the Grand Lama of Tibet, Bourke claimed, was "...dried, powdered, and sold at high prices by the priests..." who used it to "...induce sneezing, as a condiment for their food, and as a remedy for all the graver forms of illness".[98] The Assyrians offered excrement and flatus as oblations to Venus, while witches of medieval Europe kissed the anus of Satan in worship.[99] The Roman Goddess Cloacina, guardian of the cloacas, sewers and privies, Bourke remarked, may have been none other than the goddess Venus herself, since statues of Venus were often said to be found at the cloacae of Ancient Rome.

This slippage between excrement and sexual love is found in several of the ethnographies Bourke surveys. A whole chapter detailed the use of ordure in "love philters", citing as one example, a medieval French witchcraft practice, also replicated in early modern England, of feeding a cake containing "nameless ingredients" to a beloved so as to cause his infatuation.[100] In descriptions of Aztec deities, he noted, the mother of all gods Suchiquecal is said to eat *cuitlatl* [shit], though there was also a specific Aztec goddess of ordure, Tlaçolteol, mentioned in several Spanish accounts, who presided over lovers and carnal pleasures, was the goddess of "...vices and dirtiness...", of "*...basura ó pecado...*" (ordure and sin), and "...eater of filthy things".[101] The types of gods worshipped by a culture indicated a people's own past practices, Bourke claimed. Hence those whose gods were devourers of men, had once themselves been cannibals, while those whose gods presided over excrement, had most certainly once themselves been coprophagic.[102]

Other sources for Bourke's account were drawn from respectable works of late nineteenth-century ethnography and anthropology. He cited the British explorer and Arabist Sir Richard Burton's discussion of the "Holy Merde", referring to Arab uses of "khara", which Burton

[97] Ibid., 460.

[98] Ibid., 46.

[99] Ibid., 32–72.

[100] Ibid., 217, 221.

[101] Ibid., 130–131.

[102] Ibid., 129–130.

thought derived from Hindu uses of excrement as sacred medicine.[103] Bourke discussed the renowned Scottish anthropologist James Frazer, who in the highly influential 1890 work *The Golden Bough* observed that in many of the "savage" societies he described, "the conception of holiness and pollution" were "not yet differentiated…"; hence those most highly valued in society (kings, priests and chiefs) were treated with the same laws of taboo as were those deemed unclean (murderers, pregnant or menstruating women, pubescent girls, hunters).[104] Bourke thought this "inability", then, to distinguish that which had no value from that which was ultimately valuable defined the primitive psyche.

Similarly, those who failed to live as civilised subjects in modern societies might be found reverting to primitive excrement-eating: An entire chapter in Bourke's compendium was devoted to examples gathered from medical and psychiatric texts detailing coprophagy among the mentally insane. He cited the work of the Austrian neurologist Heinrich Obersteiner whose 1872 paper drew from the data of Dr A. Erlenmeyer, showing that one in every hundred insane persons "indulged in the abnormal appetite", and that "no deleterious effect on the alimentary organs was detected" in any of them.[105] Freud had worked in Obersteiner's private Vienna sanitorium prior to commencing his studies with Charcot in Paris.[106] It seems likely that he too would have known of this work by Obersteiner, since in his 1917 "On Transformation of Instinct as Exemplified in Anal Erotism'", he referred to "obsessional neurotics" in whom "regressive debasement" towards faeces was observed.[107]

There is no doubt that Freud read and was highly influenced by ethnographic works from across Europe (indeed from around the world) produced at the end of the nineteenth century. In 1913, in the same year that he published *Totem and Taboo*, Freud also wrote a preface to German translation of Bourke's *Scatalogic Rites*.[108] In this short essay he claimed as an established "fact" that "the human infant is obliged to

[103] Burton, *The Book of the Thousand Nights and a Night*, 223, 229.

[104] Frazer, *The Golden Bough*, 223.

[105] Bourke, *Scatalogic Rites of All Nations*, 29.

[106] Mayer, Sites of the Unconscious, 122.

[107] Freud, "On Transformation of Instinct as Exemplified in Anal Erotism," 131.

[108] Freud, "Geleitwort zu '*Der Unrat in Sitte, Glauben und Gewohnheitsrecht in Völker*'" (1913), 249–251.

recapitulate during the early part of his development the changes in the attitude of the human race towards excremental matters which probably had their start when *homo sapiens* first raised himself off Mother Earth", adopting the bipedal body-position.[109] This was a pivotal moment because in turning away from the olfactory stimulus of having our noses near the ground and near the anuses of other hominids, *homo sapiens* had taken an important step towards a new relationship of separation from the lower gut and its products—this was the new "organic-olfactory theory of repression", as Sulloway remarks, replacing previous views about psychosexual development as a series of defences against traumatic seduction.[110] For Freud, implicitly, the coprophagic rituals of primitive cultures were both a cause and a symptom of their inability to progress since it signified the failure to sublimate anal desire towards the goal of civilisation, and demonstrated their underdevelopment in the lack of a monetary symbolic system needed for economic development. The anal-sadistic stage of individual psychosexual development was the recapitulation of the social evolutionary "barbarous" stage, in which the boy child was confronted with the need to give up excrement in exchange for retaining the phallus but had not yet accepted this unavoidable civilisational requirement.

It is not that ideas about primitive excrement were everywhere in Freud's time, nor a particularly ubiquitous discourse. In fact, it is a striking question as to why the excremental thought of the early moderns so surprised nineteenth-century thinkers, who themselves appeared to consider the topic unspeakable. While Freud succeeded in locating particular ethnographic sources that stimulated his ideas about primitivity and the anus, descriptions of excrement were certainly not common, either in amateur ethnographic, professional anthropological, or medical texts of the nineteenth century, in the way they had been throughout the early modern period. But this was precisely how Freud's ideas tended to emerge—not via deep engagement with the central currents of intellectual disciplines such as biology and anthropology, but rather through a highly selective picking from the idiosyncratic edges of such bodies of thought. Had Freud taken anthropological knowledge seriously, he would surely have heeded Boas' warning not to assume that all cultures

[109] Freud, "Preface to Bourke's Scatological Rites of All Nations" (1913), 336.

[110] Sulloway, Freud, *Biologist of the Mind*, 377.

develop along the lines of European socio-economic history, or that such development is biologically predetermined. Even so, Freud would still have been exposed to multiple fields of conceptual meaning that related excrement to money and to civilisation, as this was a theme found in popular linguistic custom in German and English (as Freud himself considered); in debates about sewerage technologisation in European urban centres; in medical writing on constipation; and in most importantly in accounts of scatalogical history.

Bibliography

Abraham, Karl. "A Short Study of the Development of the Libido". *Selected Papers of Karl Abraham*, translated Douglas Bryan and Alix Strachey, 480–501. London: Hogarth Pressa, 1927.

Anon. *Seule réponse à l'article anonyme, sur les fosses d'aisance publié dans la Vigie de Dieppe, du 20 octobre 1863; Procès-verbal de la séance du Conseil d'Hygiène et de salubrité de Dieppe et de délégués des conseils du département de la Seine-Inférieur, présidée par M. le Sous-Préfet de Dieppe, le 18 Mai 1863*: 1–6.

Barnes, David S. *The Great Stink of Paris and the Nineteenth-Century Struggle Against Filth and Germs*. Baltimore: John Hopkins University Press, 2006.

Bernabé, K. J., C. Langendorf, N. Ford, J. B. Ronat, and R. A. Murphy, "Antimicrobial Resistance in West Africa". *International Journal of Antimicrobial Agents* S0924-8579, no. 17: 30274–30281.

Beveridge, Peter. *The Aborigines of Victoria and the Riverina as Seen by Peter Beveridge*. Melbourne: M.L. Hutchinson, 1889.

Boas, Franz. *The Mind of Primitive Man*. New York: Macmillan, 1911.

Bölsche, Wilhelm. *Das Liebesleben in der Natur: Eine Entwicklungsgeschichte der Liebe*, 3 Bd. Florence/Leipzig: Diederichs, 1893–1903.

Bourke, John G. *Compilation of Notes and Memoranda Bearing Upon the Use of Human Ordure and Human Urine in Rites of a Religious or Semi-Religious Character Among Various Nations*. Washington: U.S. War Department, 1888.

Boyle, Robert. "The Usefulness of Experimental Philosophy, By Way of Exhortation to the Study of It". In *The Philosophical Works of the Honourable Robert Boyle*, vol. 1, edited by P. Shaw, 3–181. London: W. Innys and R. Manby, 1738.

Boyle, Robert. "The Nature, Properties, and Effects of Effluvia". In *The Philosophical Works of the Honourable Robert Boyle*, vol. 1, edited by P. Shaw, 403–438. London: W. Innys and R. Manby, 1738.

Buckle, Henry Thomas. *Miscellaneous and Posthumous Works*, vol. II, edited by Helen Taylor. London: Longman Green, 1872.

Burke, Peter. *What Is Cultural History?* Cambridge: Polity, 2008.

Burton, Richard F. *The Book of the Thousand Nights and a Night, A Plain and Literal Translation of the Arabian Nights Entertainments*, vol. 2, translated and annotated by Richard F. Burton. London: Burton Club, 1885.

Carden-Coyne, Ana and Christopher Forth, eds. *Cultures of the Abdomen: Dietetics, Digestion and Obesity in the Modern World.* New York: Palgrave, 2005.

Chalmers, James. "Remarkable Constipation. Inability to Empty the Bowels During Three Years". *The London Medical Gazette, New Series* 1, 1842–1843: 20–21.

Choqart, C. *Aux consommateurs de chocolat et de thé: histoire d ces deux aliments, leurs propriétés hygiéniques, leur fabrication et leur commerce; recettes les meilleures pour les préparer.* Paris: Chocolaterie Impériale, 1868.

Considérant, Victor, *Description du phalanstère et considérations sociales sur l'architechtonique.* Paris: Librairie sociétaire, 1838.

Corbin, Alain. *Le Miasme et la jonquille: l'odorat et l'imaginaire social, xviiie-xixe siècles.* Paris: Editions Aubier Montaigne, 1982.

Cotti, Patricia, "Sexuality and Psychoanalytic Aggrandisement". *History of Psychiatry* 22 (2011): 58–74.

Delaporte, François. *Le Savoir de la maladie: essai sur le cholera de 1832 à Paris.* Paris: Presses Universitaires de France, 1990.

Douglas, Mary, *Purity and Danger: An Analysis of Concepts of Pollution and Taboo.* London: Routledge & Kegan Paul, 1966.

Dundes, Alan. *Life Is Like a Chicken Coop Ladder: A Study of German National Character Through Its Folklore.* Detroit: Wayne State University Press, 1989.

Fauvel, Pierre. "Action du chocolat et du café sur l'excrétion urique". *Extrait des comptes rendues des séances de la Société de Biologie*, Séance du 16 Mai, 1908, Tome LXIV: 854.

Ferenczi, Sandor. "The Ontogenesis of the Interest in Money". In *Contributions to Psychoanalysis*, edited by Ernest Jones. Boston: Richard Badger, 1916.

Franklin, Alfred. *La Vie privée d'autrefois, Tome VIII. L'Hygiène; Arts et métiers, modes, mœurs, usages des Parisiens du douzième au dix-huitième siècle d'après des documents originaux ou inédits.* Paris: Librairie Plon, 1890.

Frazer, James G. *The Golden Bough: A Study in Magic and Religion*, vol. 1. New York: Macmillan, 1947.

Freud, Sigmund. "Geleitwort zu 'Der Unrat in Sitte, Brauch, glauben und Gewohnheitsrecht der Völker von John Gregory Bourke". In *Gesammelte Schriften* 11, 249–251. Leipzig: Internationaler Psychoanalytischer Verlag, 1928.

Freud, Sigmund. *Totem and Taboo: Resemblances Between the Psychic Life of Savages and Neurotics,* translated by James Strachey. London: Routledge, 1950.

Freud, Sigmund. "On the Transformation of Instincts with Special Reference to Anal Erotism". In *Collected Papers*, Issue 8, translated by Joan Rivière and Alix Strachey, 164–171. New York: Basic Books, 1959.
Freud, Sigmund. *Civilisation and Its Discontents,* translated by James Strachey. New York: W. W. Norton, 1962.
Freud, Sigmund. "Character and Anal Erotism". In *The Freud Reader*, edited by Peter Gay, 294–297. London: Vintage, 1989.
Freud, Sigmund. "The Dissolution of the Oedipus Complex". In *The Freud Reader*, edited by Peter Gay, 661–664. London: Vintage, 1989.
Freud, Sigmund. "The Claims of Psychoanalysis to Scientific Interest". In *The Standard Edition of the Complete Works of Sigmund Freud*, vol. 13, edited by James Strachey, 165–192. London: Hogarth Press, 1994.
Freud, Sigmund. "Preface to Bourke's Scatalogic Rites of All Nations (1913)". In *The Standard Edition of the Complete Works of Sigmund Freud*, vol. 12, edited by James Strachey, 333–339. London: Vintage, 2001.
Frexinos, Jacques. *Voyage sans transit: Géographie mondiale de la constipation.* Paris: Editions Médigone pour les Laboratoires Beaufour, 1997.
Gay, Peter. *Reading Freud: Explorations and Entertainments.* New Haven and London: Yale University Press, 1990.
Gay, Peter. *The Cultivation of Hatred: The Bourgeois Experience, Victoria to Freud.* New York and London: W. W. Norton, 1993.
Gay, Peter. *Freud, A Life for Our Time.* New York: W. W. Norton, 2006.
Gerrand, Roger-Henri. *Les Lieux: Histoire des commodités.* Paris: La Découverte, 1997.
Glick, Thomas F. *The Comparative Reception of Darwinism.* Chicago: University of Chicago Press, 1988.
Gould, Stephen J. *Ontogeny and Phylogeny.* Cambridge, MA: Harvard University Press, 1977.
Guinane, Caitriona M., and Paul D. Cotter. "The Role of the Gut Microbiota in Health and Chronic Gastrointestinal Disease: Understanding a Hidden Metabolic Organ". *Therapeutic Advances in Gastroenterolog* 6, no. 4 (2013): 295–308.
Haeckel, Ernst. *The Riddle of the Universe* [*Die Weltraetsel*, 1899]. Buffalo, NY: Prometheus Books, 1992.
Hallé, M. *Recherches sur la nature et les effets du méphitisme des fosses d'aisance.* Paris: L'Imprimerie de Ph.-D/Pierres, 1785.
Hau, Michael. *The Cult of Health and Beauty in Germany, 1890–1930.* Chicago: Chicago University Press, 2003.
Henry, Édouard-Georges. *Hygiène colonial.* Lorient : A. De La Morinière, 1893.
Herbert, Christopher. "Filthy Lucre: Victorian Ideas of Money". *Victorian Studies* (Winter 2002): 185–213.
Himmelfarb, Gertude. "The "Real" Marx". *Commentary* 79, no. 4 (1985): 39.

Ho Bak, Seon, Hyun Ho Choi, Jinhee Lee, Mi Hee Kim, Youn Hee Lee, Jin Su Kim and Young-Seok Cho. "Fecal Microbiota Transplantation for Refractory Crohn's Disease". *Intestinal Research* 15, no. 2 (2017): 244–248.

Hornibrook, Frederick Arthur. *Culture of the Abdomen: The Cure of Obesity and Constipation*. Chicago: W. Wood & Co, 1924.

Hugo, Victor. *Les Miserables*, dans *Oeuvres Complètes*, vol. 11, edited by Jean Massin. Paris: Club Français du Livre, 1969.

Inwood, Stephen. *A History of London*. London: Papermac, 2000.

Khanna, Ranjanna. *Dark Continents; Psychoanalysis and Colonialism*. Durham and London: Duke University Press, 2003.

Koselleck, Reinhart. *The Practice of Conceptual History: Timing History, Spacing Concepts*, translated by Todd Presner, Kerten Behnke and Kobst Welge. Stanford: Stanford University Press, 2002.

Kristeva, Julia. *Pouvoirs de l'horreur: Essai sur l'abjection*. Paris: Éditions du Seuil, 1980.

LaGuardia, David. "Doctor Rabelais and the Medicine of Scatology". In *Fecal Matters in Early Modern Literature and Art: Studies in Scatology*, edited by Jeffrey Persels and Ganim Russel, 38–54. New York: Routledge, 2004.

Laporte, Dominique. *L'Histoire de la merde*. Paris: C. Bourgois, 1993.

Le Cade, Docteur Amable. "Cure remarquable d'une constipation de 40 jours en 10 minutes par l'electro-thérapie indutive". *Lyon Médical*, Deuxième année, Tome IV (1870): 242–244.

Le Page, G, L. Gunnarsson, J. Snape, and C. R. Tyler. "Integrating Human and Environmental Health in Antibiotic Risk Assessment: A Critical Analysis of Protection Goals, Species Sensitivity and Antimicrobial Resistance". *Environment International* S0160-4120, no. 17: 30900–30905.

Leach, Jeff. "Gut Microbiota: Please Pass the Microbes". *Nature* (Correspondance) 504. 33 (December 2013): 1.

Lehrer, Steven. "Duodenal Infusion of Feces for Recurrent Clostridium Difficile". *New England Journal of Medicine* 368, no. 22 (May 2013): 2144.

Lopez, Joanna, and Ari Grinspan. "Fecal Microbiota Transplantation for Inflammatory Bowel Disease". *Gastroenterology & Hepatology* 12, no. 6 (2016): 374–380.

Lloyd-Price, Jason, Galeb Abu-Ali and Curtis Huttenhower. "The Healthy Human Microbiome". *Genome Medicine* 8, no. 51 (2016): 1–11.

Mayer, Alfred. "La Canalisation souterraine de Paris". In *Guide-Paris, par les principaux écrivains et artistes de la France*, edited by Corinne Verdet, 184. Paris: La Découverte and Maspero, 1983.

Mayer, Andreas. *Sites of the Unconscious: Hypnosis and the Emergence of the Psychoanalytic Setting*. Chicago: University of Chicago Press, 2013.

Mayhew, Henry. *London Labour and the London Poor*, 4 vols, vol. 2. New York: Dover Publications, 1968.

McClintock, Anne. *Imperial Leather: Race, Gender, and Sexuality in the Colonial Contest.* New York: Routledge, 1995.
Moore, Alison M. *Sexual Myths of Modernity: Sadism, Masochism and Historical Teleology.* Lanham: Lexington Books, 2015.
Moore, Alison M. "Historicizing the Modern European Excremental World-View". *Advanced Research in Gastroenterology & Hepatology* 10, no. 1 (June 2018).
Neu, Jerome, ed. *The Cambridge Companion to Freud.* Cambridge: Cambridge University Press, 1991.
Parent-Duchâtelet, Alexendre Jean-Baptiste. *De la prostitution dans la ville de Paris, considérée sous le rapport de l'hygiène publique, de la morale et de l'administration : Ouvrage appuyé de documents statistiques puisés dans les archives de la Préfecture de police, avec cartes et tableaux.* Paris: J.B. Baillière, 1837. Tome 1.
Payen, Jean-François. *Essai sur les eaux minérales, thermales de Louesche.* Paris: Blaise jeune, 1828.
Payen, Jean-François. *Notice bibliographique sur Montaigne.* Paris: E. Duverger, 1837.
Reclus, Elie. *Le primitif d'Australie: Les Non-non et les Oui-oui, étude d'ethnologie comparée.* Paris: E. Dentu, 1894.
Reid, Donald. *Paris Sewers and Sewermen: Realities and Representations.* Cambridge, MA: Harvard University Press, 1991.
Rémy Amouroux. "'Le Précieux livre de W. Bölsche': Freud at la culture évolutionniste allemande du début du vingtième siècle". *Gesnerus* 61 (2004): 24–36.
Schmidt, Joseph and Mary Simon. "Holy and Unholy Shit: The Pragmatic Context of Scatological Curses in Ealry Modern Gmeran Reformation Satire". In *Fecal Matters in Early Modern Literature and Art: Studies in Scatology*, edited by Jeffrey Persels and Ganim Russell, 109–119. New York: Routledge, 2004.
Schorr, S. L., M. Candela, S. Rampelli, M. Centanni, C. Consolandi, G. Basaglia, S. Turroni, E. Biagi, C. Peano, M. Severgnini, J. Fiori, R. Gotti, G. De Bellis, D. Luiselli, P. Brigidi, A. Marbulla, F. Marlowe, A. G. Henry, A. N. Crittenden. "Gut Microbiome of the Hadza Hunter-Gatherers". *Nature Communications* 15, no. 5 (2014): 3654. https://doi.org/10.1038/ncomms4654.
Schurig, Martin. *Chylologia historico-medica.* Dresden: J.C. Zimmermanni et J.N. Gerlachii, 1725.
Strachey, James. "Editor's Note". In *The Standard Edition of the Complete Psychological Works of Sigmund Freud*, Volume VII (1901–1905), translated by James Strachey, 3–6. London: Vintage Books, 2001.
Sulloway, Frank. *Freud, Biologist of the Mind: Beyond the Psychoanalytic Legend.* New York: Basic Books, 1979.
Van der Geest, Jacques. "Not Knowing about Defecation". In *Of Knowing and Not Knowing in the Anthropology of Medicine*, edited by Roland Littlewood, 75–86. London and New York: Routledge, 2007.

van Hecke, O, K. Wang, J. J. Lee, N. W. Roberts, and C. C. Butler. "Implications for Antibiotic Resistance for Patients' Recovery from Common Infections in the Community: A Systemic Review and Meta-Analysis". *Clinical Infectious Diseases* 65, no. 3: 371–382.

Walton, Jean. *Fair Sex, Savage Dreams: Race, Psychoanalysis, Sexual Difference*. Durham: Duke University Press, 2001.

Whatmore, Richard. *What is Intellectual History?* Cambridge: Polity, 2016.

Whorton, James C. *Inner Hygiene: Constipation and the Pursuit of Health in Modern Society*. New York: Oxford University Press, 2000.

Zweiniger-Bargielowska, Ina. "The Culture of the Abdomen: Obesity and Reducing in Britain, circa 1900–1939". *Journal of British Studies* 44, no. 2 (April 2005): 239–274.

CHAPTER 5

Food for Thought: Consuming and Digesting as Political Metaphor in French Satirical Prints

Dorothy Johnson

Political satire, and its manifestation in visual culture in the form of caricatural prints, is a time-honored mode of undercutting and ridiculing rulers, the ruling class, and political enemies through humorous critique. Satirical political prints, because of potential widespread dissemination and appeal based on a direct visual language that could be understood and appreciated by a broad, often illiterate public, have served throughout history to allay political fears and anxiety by revealing the stupidity, cupidity, and corruption of governments.[1] Seriocomic images of the digestive process used to satirize the body politic typically appear at

[1] Reichardt and Kohle, *Visualizing the Revolution*; Hunt, "La psychologie politique des caricatures révolutionnaires," 33–42.

D. Johnson (✉)
University of Iowa, Iowa City, IA, USA

M. Mathias and A. M. Moore (eds.), *Gut Feeling and Digestive Health in Nineteenth-Century Literature, History and Culture*, Palgrave Studies in Literature, Science and Medicine,
https://doi.org/10.1007/978-3-030-01857-3_5

moments of intense political anxiety, crisis, and polarization.[2] This was true in France in the late eighteenth and early nineteenth centuries.

Images of consuming, digesting, and excreting used metaphorically were widespread during agonistic moments of political instability and conflict in France, such as the period of the first French Revolution of 1789. The second French Revolution of 1830 and the turbulent early years of the July Monarchy witnessed a resurgence of the theme in political caricatures. The present article examines selected examples from each of these periods of political lability that serve as case studies for the metaphorical uses of consuming and digesting as political critique. The examples chosen appeared in a period when art and anatomy were confluent, when medicine and the arts were closely allied, and when physiologists were investigating the emotional and psychological manifestations of indigestion, of what was happening in the guts, or "gut feeling". The focus on digestion in late eighteenth-century medical circles, in concomitance with visual and literary sources, plays a role in helping us understand the levels of meaning in French political caricatures in which consuming and digesting were paramount.

The mid to late eighteenth century in France witnessed an intensified fascination with visualizing the interior body. This was due in large measure to the need in the medical and biological sciences for detailed visual information in two- and three-dimensional formats about the organs and viscera and how they fit together, information based on dissection. Peering into the body via dissection in order to better understand its inner biological life had become an obsession in late Enlightenment medicine, particularly in the Vitalist School of Medicine in Montpellier.[3] This period witnessed an efflorescence of illustrated anatomy books and three-dimensional wax models for surgeons as well as for artists, who, by 1794 had a new requirement in academic art training—not just the observation of dissection but its practice as well.[4] Artist/illustrators who produced anatomy books or wax anatomical models transformed the blood and guts of the dissected cadaver into an ordered, bloodless materialization of the inside of the body.

[2]For a history of prints satirizing the body politic see Baridon and Guédron, *L'Art et l'histoire de la caricature*.

[3]Williams, *A Cultural History of Medical Vitalism* and *The Physical and the Moral*. See also Vila, *Enlightenment and Pathology* and Stafford, *Body Criticism*.

[4]Dorothy Johnson, "Anatomie, réalité, idéalité," and "The Body Speaks," 29–62.

Dissection and its visualization in two- and three-dimensional replicas were of great use for doctors and surgeons and helped to inaugurate modern medicine. The fascination for artists of seeing representations of organs and viscera and all that was hidden beneath the seal of the skin appealed to visual as well as intellectual or scientific curiosity, a curiosity, for example, that had inspired Leonardo da Vinci's anatomical drawings (Leonardo remained a prepotent example for the instantiation of the artist-anatomist in this period). Understanding the anatomy of bones and muscles could serve artists well in attempts to replicate the human figure. Observing first-hand brains, hearts, livers, stomachs, and intestines, was another matter. Throughout most of the eighteenth century too much exposure to anatomical visualizations was considered a threat to artists whose goal was considered to be ideal beauty. The *locus classicus* for the expression of this apprehension came from Diderot who had written about his concerns regarding the study of anatomy in his 1766 *Essais sur la peinture* (*Notes on Painting*):

> Study of the *écorché* doubtless has its advantages; but is it not to be feared that this *écorché* might remain in the imagination forever; that this knowledge might encourage the artist to become enamored of his knowledge and show it off; that his vision might be corrupted, precluding attentive scrutiny of surfaces; that despite the presence of skin and fat, he might come to perceive nothing but muscles, their beginnings, attachments, and insertions.[5]

The *Académie royale de peinture et de sculpture* was in complete consonance with these sentiments.[6] It was only in the later years of the eighteenth century in France and especially after the Revolution that anatomical study for artists took on renewed importance and prestige.[7]

During the eighteenth century, physiologists also intensified investigations of the internal biological processes that were invisible. Among

[5] Diderot, "Notes on Painting," 193. "L'étude de l'écorché a sans doute ses avantages; mais n'est-il pas à craindre que cet écorché ne reste perpetuellement dans l'imagination; que l'artiste n'en devienne entêté de la vanité de se montrer savant; que son oeil corrompu ne puisse plus s'arrêter à la superficie; qu'en dépit de la peau et des graisses, il n'entrevoie toujours les muscles, son origine, son attache et son insertion". Diderot, *Oeuvres complètes*, vol. 10, 463–464.

[6] Johnson, "The Body Speaks," 35–39.

[7] Johnson, "Anatomie, réalité, idéalité".

the most complex and least understood of these processes was digestion. Digestive problems and symptoms had been a subject of humor in art and literature going back to antiquity, and remedies had been of keen interest to every person who had suffered indigestion and the doctors who treated them. The belief in the biological origins of temperaments still prevailed in the eighteenth century with the choleric temperament, influenced by too much yellow bile, the one most closely associated with the digestive process.[8] In the eighteenth century, the digestive process became a focus of medical investigations by the Vitalist School in Montpellier whose adherents sought to understand all aspects of human biology and the role played by physiology in behavior. In a seminal article on the topic, Elizabeth Williams[9] writes of Vitalist physicians: "thinking about digestion was linked to a broad range of issues: how to differentiate matter, spirit, and life; how to discern the boundaries of the body and psyche, and, perhaps of greatest moment to contemporaries, how to ensure health and ward off illness".[10] Williams contests the notion that the historical study of ideas about digestion belongs solely to our understanding of human biology:

> An activity governed by involuntary, unconscious mechanisms, digestion would seem to have no history at all but to form part of the basic biological equipment of the species. Only recently have historians even thought to ask whether individual body parts or functions can be said to have, in some sense, a history or culture.[11]

If we think in terms of the realm of depictions of the human body in visual culture, we can see that such depictions are ineluctably filiated with prevailing cultural and historical ideas and ideals of the body. Williams discusses how Vitalist physiologists were investigating the entire digestive cycle, from the consuming of food to the various processes it undergoes as it travels through the body and how food becomes transmuted in the process to serve as the basis of vital energy. The article on digestion written by the Vitalist physician, G. F. Venel for the *Encyclopédie* of Diderot and d'Alembert, provides a good example of Enlightenment

[8] Halliday, *The Temperamental Nude.*

[9] Williams, "Food and Feeling," 203–221.

[10] Ibid., 204.

[11] Ibid.

knowledge and debates about the complex digestive process.[12] Williams also discusses J.-C.-M.-G. Grimaud, who emphasized the link between digestion and psychology in his treatises on nutrition in the 1780s.[13] The extent to which medical interest in and knowledge of digestion continued to intensify can be seen in the comprehensive 60-volume *Dictionnaire des sciences médicales* of 1812–1818: Almost 100 pages are devoted to the digestive processes.[14]

Artists, like the educated public, were likely well aware of the interest in digestion and its impact on human emotion and behavior. I have studied elsewhere the profound intersections of art with anatomy and the medical sciences in France circa 1800.[15] While we need to keep in mind the importance of the natural and medical sciences, which were held in very high esteem in late eighteenth-century France, in any representations of the digestive process used metaphorically, artists were keenly attentive to visual precedents in the depictions of consuming, digesting, and excreting. A rich array of multifarious imagery, especially in terms of the representation of digestive disorders, can be found in medieval, Renaissance, Baroque, and eighteenth-century art. Seriocomic descriptions of indigestion, particularly violent expulsions of digestive matter from the body, also abound in historical and contemporaneous writings that were very popular in France at the end of the eighteenth century: Rabelais, Swift, Jacques Hébert, and the Marquis de Sade immediately come to mind.[16] Several of these writers will be discussed as possible sources of inspiration for the caricatures we will investigate in this study. While literary and visual sources are often rebarbative, crude, and comical, they typically have serious metaphorical and political meanings to convey.

During the early years of the French Revolution, digestive metaphors were used in caricatures that conveyed scabrous attacks on the king and the aristocracy. A remarkable image of this type is found in a large, anonymous engraving from around 1791, *Le Ci-Devant Grand Couvert de*

[12] Ibid., 204–205.

[13] Ibid., 208–212.

[14] "Digestion," in *Dictionnaire des sciences Médicales*, vol. IX, 354–451.

[15] Johnson, "Anatomie, réalité, idéalité," "The Body Speaks".

[16] Boime has discussed the importance of these authors vis-à-vis late eighteenth-century scatological prints in "Jacques-Louis David," 72–81. Reprinted with minor modifications in Cuno, *Politique et Polémique*, 69–86

Fig. 5.1 Anonymous, *Le Ci-Devant Grand Couvert de Gargantua moderne en famille*, 1791, engraving, Bibliothèque Nationale de France, Paris

Gargantua moderne en famille (*How the Modern Gargantua and His Family Once Dined*, Fig. 5.1).[17] Louis XVI as Rabelais' gluttonous giant is seated at table with nobles to our right and Marie Antoinette on the left. Rabelais' *Gargantua and Pantagruel* was popular in the late eighteenth century, and in the nineteenth century the giants would be beloved by the Romantics, including Victor Hugo.[18] In the engraving, the small-scale Lilliputian-type subjects (recalling Swift) of the modern Gargantua's realm bring food and wine from all the regions of the land and sea and carry them up on ladders to his table. Louis XVI is in the process of bringing a whole pig speared on a fork to his mouth. In his seminal study, the great

[17] Cuno, Ibid.

[18] Bakhtin, *Rabelais and His World*, 2–18, 43. Doré's illustrations to the 1873 edition reveal the continued popularity of the book.

Rabelais scholar, Bakhtin, discussed in detail the episodes in *Gargantua and Pantagruel* that describe the consumption by the giants of vast amounts of food and drink that resulted in mammoth excretions that could destroy entire towns, and the role of the gaping, consuming mouth in imagining the grotesque body.[19] Rabelais devotes long passages to what happens when food enters the mouth and travels to the bowels below (Pantagruel's mouth, for example, encompassed twenty-five kingdoms and when he suffered a bad bout of indigestion teams of human laborers had to be lowered into his stomach for excavation and repair).[20] In the engraving of the modern Gargantua, while the king happily devours and consumes, his spouse, Marie-Antoinette, fills an enormous glass with the blood spouting from a hapless subject whose throat has been slit by the notorious royalist General, the Marquis de Bouillé, nicknamed the "Butcher of Nancy" for his violent and bloody suppression of a military uprising in Nancy on August 31, 1790 that resulted in 500 civilians and soldiers being killed. The caption informs us that Marie Antoinette wonders if there will be enough blood in the glass for Louis XVI to bathe in. Vampirism, the bloodbath, and the voracious consumption of all that the country produced as food, satirize the insatiable appetites of the King, his family, and court who devour the country and its subjects.

That the drinking of blood was associated with cannibalism is evinced in a print that depicts the horrors of the Terror, especially a blood thirst associated with the guillotine. *Les formes acerbes* (*Harsh Manners*, Fig. 5.2), commemorates the execution of Joseph Le Bon, a disciple of Robespierre, on October 15, 1795. The print pictures the madness of Le Bon as he drinks the blood of his guillotined victims. An allegorical vision of Justice unveiling Truth to the National Convention floats above two guillotines. To the left are hapless prisoners, future victims of Le Bon, pleading for justice. Below Justice and Truth stands the agent of the Terror, the "cannibal" Le Bon, who is described in the caption in the following terms:

> This allegorical engraving represents Joseph Le Bon, positioned between the two guillotines of Arras and Cambray, holding two chalices. In one he receives and in the other drinks the blood of numerous victims, above

[19] Bakhtin describes the gaping mouth of the grotesque body as the "open gate leading downward to the bodily underworld". Ibid., 325–339.

[20] Ibid., 337–338.

Fig. 5.2 *Les formes acerbes*, etching attributed to a drawing of Lafitte after Poirier de Dunkerque, 1795, Bibliothèque nationale de France, Paris

> 550, immolated in the two communities. He stands on groups of cadavers that are piled on top of one another. On one side two furies, worthy companions of this cannibal, incite the animals that are less ferocious than they are to devour the remains of the unfortunates that they can no longer torment ....[21]

[21] "Cette gravure allégorique représente Joseph Le Bon, posté entre les deux guillotines d'Arras et de Cambray tenant deux calices dans lesquels il reçoit d'une main et s'abreuve de l'autre du sang de ses nombreuses victimes, immoleés au dela de 550 dans les deux communes. Il est monté sur des groupes des cadavres entassés les uns sur les autres; d'un côté deux furies dignes compagnes de ce cannibal animent des animaux moins féroces qu'elles à dévorer les restes des malheureuses qu'elles ne peuvent plus tormenter". Discussed in Johnson, "Visceral Visions".

Fig. 5.3 Anonymous, *Un monstre à trois têtes désignant les trois états de l'aristocratie*, colored etching, 1790, British Museum, Trustees of the British Museum

The engraving refers to historical accounts of the overzealous use of the guillotine in the provinces and makes explicit in brutal and gory visual terms, the literal bloodthirstiness of Le Bon who drinks with great zeal the blood pouring from the decapitated victims he had sent to the guillotine, while below him savage animals consume the bodily remains.

Cannibalism in a broad sense was used metaphorically by Revolutionaries and counter-Revolutionaries as a means of showing how one group consumed and digested another. An anti-royalist caricature from this period, for example, uses cannibalism as a metaphor for the aristocracy's devouring of the people, "Three Heads of the Aristocracy" (Fig. 5.3).[22] The 1790 colored etching depicts a skeletal chimera, a hydra with three grotesque heads of the aristocracy that has swallowed the cadaver of the "people" seen in its bowels. The caption reads: "A monster with three heads designating the three states of the aristocracy is busy devouring the remains of the cadaver of the people that it

[22] Ibid.

has mercilessly swallowed in its carnivorous bowels". The caption also describes the crawling figure of a monk (Fanaticism) with a bifurcated snake's tail. A nun (Hypocrisy) is astride the monk's back and holds a snake that secretes aristocratic poison. It seems that the cadaver of the "people", encased in the monster's bowels, is indigestible. The meal consumed by the aristocracy, the "people", will never provide digestible material to put flesh on the chimera's bones.[23]

The metaphor of cannibalism would also inflect counter-Revolutionary prints. One of the most famous is James Gillray's 1792 acerbic colored etching, "Un petit souper à la Parisienne;—or—a Family of Sans-culottes refreshing, After the Fatigues of the Day" (Fig. 5.4).[24] In this brutal caricature, a cannibal family of *sans-culottes* (people of the working class and the poor who became radicalized in the Revolution), including ravenous children and a crone/grandmother, voraciously dines on the dismembered bodies of the French aristocracy, parts of which also serve as chairs for this barbaric meal. The political message is of signal clarity. This etching forms part of a series of Gillray's counter-Revolutionary prints. Gillray's caricatures of the cannibal *sans-culottes* bring to mind a horrific episode of cannibalism that was reported in the village of Poleymieux near Lyon on June 26, 1791, just two days after Louis XVI's ill-fated flight to Varennes.[25] Guillin du Montet, from a Lyon family of magistrates, had returned home from Senegal where he became rich dealing with the slave trade. He purchased the seigneury of Poleymieux and proved to be a violent, avaricious, and abusive Lord of the territory whose actions enraged the village inhabitants. After the abolition of feudal rights by the constituent assembly on August 4, 1789, Guillin du Montet's days were numbered. An angry mob invaded his chateau on June 26, 1791, killed and dismembered him. Based on accounts by his widow and several eye-witnesses, the local butcher cut him to pieces and threw his body fragments on a type of celebratory bonfire. Two young men retrieved his right hand and arm and his heart and "bit into the pieces of flesh" and "kissed them" as they shouted verbal abuses at the hated aristocrat they were consuming. His widow

[23] Ibid.

[24] Boime, "Jacques-Louis David," 76–77.

[25] This episode was written about extensively by Viola, "The Rites of Cannibalism," 157–174.

Fig. 5.4 James Gillray, *Un petit souper, à la Parisienne;—or—A Family of Sans-culottes Refreshing After the Fatigues of the Day*, colored etching, 1792, British Museum, Trustees of the British Museum

referred to them as "cannibals" who "roast their victim's limbs and then devour them".[26]

Cannibalism persisted as a metaphor in French counter-Revolutionary caricatures from the period of the Terror. We see a victim being roasted on a spit in the 1793 *The Devil's Supper* (*Souper du diable*), an anonymous engraving that recalls the images of devils' roasting heretics from the medieval Mystery plays.[27] A devil eagerly prepares to carve up his

[26] Ibid., 166–169.

[27] Cuno, *Politique et Polémique*, 227–228, Fig. 118. Bakhtin discusses the role of devils in Mystery plays in *Rabelais*, 347–348.

meal: A fully clothed Marat who is roasting on a spit. A fellow devil undresses a young woman sans-culotte whom he will feast on. In another engraving, *The Devil's Indigestion* (*L'indigestion du diable*), also from 1793, a devil, seated on a *chaise percée*, a type of toilet seat, has suffered such great indigestion from consuming *sans-culottes* that he is expelling them by vomiting and defecating simultaneously, a type of severe colic.[28] The monsters who are helping him get through this horrible episode of indigestion exclaim: "We told you that nothing is more indigestible than all these scoundrels".[29]

Perhaps recalling the cannibalistic acts during the Revolution, the philosophe and Encyclopédiste, the Abbé Morellet, proposed in Swiftian fashion (referencing *A Modest Proposal*) that cannibalism could be a means of feeding the nation under Jacobin rule.[30] In his 1794 *Nouveau moyen de subsistence pour la nation, proposé au Comité de salut public en messidor de l'an II* ("New means of subsistence for the Nation proposed to the Committee of Public Safety during Messidor, Year II"), he states:

> I propose to those patriots who butcher their fellow men that they should eat the flesh of their victims, and, given the state of famine to which they have reduced the nation, should feed those that they allow to live on the flesh of those that they kill. I would even propose that a *national butchery* be established in accordance with the designs of the great artist and patriot David, and that a law be passed to oblige all citizens to buy supplies there at least once a week, under pain of being imprisoned, deported or having their throats cut as suspects, and [I] demand that at every patriotic festival there should be a course of this kind, which would be the true communion of patriots, the Eucharist of the Jacobins.[31]

In this sardonic passage, Morellet conveys the cannibalistic as well as the satanic nature of the Jacobins who butcher those they condemn for opposing the Revolution and feed them to the "patriots" as a form of the communion or Eucharist in something akin to a black mass (the Jacobin patriotic festivals).

[28] Cuno, ibid., 227–228, Fig. 119.

[29] "Nous vous l'avons bien dit, rien d'aussi indigestes que tous ces coquins".

[30] Swift was much admired by the philosophes such as Voltaire and his 1729 "Modest Proposal…" had been translated into French in 1733. See "Swift's First Voyages," 5–16. See also Lee, *Swift's Scatological Satire*.

[31] Cited in Arasse, *La guillotine*, 162, no. 61.

The metaphor of cannibalism and the performative act itself was just one mode of using digestive processes as a form of conveying how opposing political groups could consume and digest one another, recalling the powerful line from King Lear: "Humanity must, perforce, prey upon itself, like monsters of the deep" (Act IV, scene 2). Another mode of acerbic attack was scatological imagery. Revolutionary caricatures, especially those that targeted a mass audience, including the *sans-culottes*, often used scatological imagery to get moral and political metaphors across in a direct way. In 1792 an anonymous French engraver created a scathing scatological image of digestive explosion and excretion as political metaphor, *Bombardment of All the Thrones of Europe and the Fall of Tyrants for the Good of the Universe* (*Bombardement de tous les trônes de l'Europe et la chute des tyrans pour le bonheur de l'univers*).[32] In this metaphorical political purge, an allegorical figure of Liberty sets the fuse on a cannon used to give Louis XVI an enema to cure his "indigestion". The King vomits out arrows identified as "vetos" that fall on the monarchs of Europe, with Catherine the Great of Russia standing tall above the others (she had tried unsuccessfully to lead the European monarchs to attack Revolutionary France). The British Prime Minister William Pitt and the Pope are part of the group being bombarded. A huge Prussian eagle carries a crown to use as an umbrella to no effect. At the same time as the king's vomit, the royal crowd are bombarded by projectile liquid excretions from the exposed bottoms (literally "sans culottes") of the kneeling members of the National Assembly who form the base on which Liberty stands. Written on the bands of liquid are the words, "Liberté ca ira". One of the kings who receives the weaponized vomit and feces exclaims: "Helas! Helas!. This indigestion of the French will cost me my crown".[33] One year later Gillray created a scatological print that shows England bombarding back, *The French Invasion: A New Map of England; or John Bull Bombarding the Bumboats*.[34] The map of England is formed by the squatting figure of George III who defecates on a fleet of ships leaving France's northern coast to invade England. The map of France is the head of a *sans-culotte*. Gillray uses scatological humor to allay British fears of a French invasion similar to the metaphorical imagery used in "Bombardement de tous les trônes".

[32] Cuno, *Politique et polémique*, 219–220, Fig. 110.

[33] Hélas, hélas, "Cette maladie des français me coutera ma couronne".

[34] Boime, "*Jacques Louis David*," 78, Fig. 10.

Scatological imagery used as a means of one group debasing another had a long history going back to antiquity. Bakhtin discussed its metaphorical uses in Rabelais as well as its sources in the festivals, theater, and rituals of fairs and carnivals in medieval and Renaissance France.[35] One group flinging feces at another had great metaphorical force and Bakhtin describes how verbal abuses using excrement coincided with actual performative acts.[36] Scatological imagery as a form of debasement re-emerged in popular French satirical imagery during the Revolution as Albert Boime has demonstrated.[37] Swift's *Gulliver's Travels*, particularly the episode where the Yahoos climb up into the trees and drop their excrement on Gulliver, inflected both British and French prints in the earlier 1790s, as evinced in the "bombardment" caricatures discussed above.[38] Boime discusses this type of imagery in the context of the Jacobin popularist writer, Jacques Hébert, who assumed the narrative persona of a sans-culotte, Père Duchesne, in his journal of the same name. The journal, popular from 1790 to 1794, when Hébert was guillotined, is replete with terms of abuse directed toward the aristocracy, using slang terms of consuming, digesting, and excreting.[39]

During the height of the popularity of Hébert's *Père Duchesne*, the great history painter, Jacques-Louis David, was called upon to create pro-Revolutionary caricatures. The leader of the French School who at the behest of the government organized Revolutionary festivals and funerals, and who served as deputy to the Convention from 1792 to 1794, produced two scatological caricatures, the most anomalous works of his career: "The English Government" ("Le Government Anglois", Fig. 5.5) and "The Royal Army of Jugs" ("L'Armée royale des cruches", Fig. 5.6).[40] In 1793, Carnot, head of the *Comité du salut public* (The Committee of Public Safety), had commissioned David to use his great talents to create caricatures that would awaken the public to the "ridiculous and atrocious" enemies of the Republic (namely, England), and that would serve a didactic purpose—to help "the people" understand

[35] Bakhtin, *Rabelais*, 147–176.

[36] Ibid., 148–149.

[37] Boime, "Jacques-Louis David," 78–79.

[38] Swift, *Gulliver's Travels*, Part IV, 258–259.

[39] These were studied in detail by Boime, "Jacques-Louis David" and Cuno, "En temps de guerre," 521–542.

[40] Boime, ibid., 73–75.

Fig. 5.5 Jacques-Louis David, *The English Government*, 1794, engraving, Bibliothèque Nationale de France, Paris

Fig. 5.6 Jacques-Louis David, *The Royal Army of Jugs*, 1794, engraving, Bibliothèque Nationale de France, Paris

the dangers of foreign resistance to the Revolution. During the year he took to complete the compositions for the two engravings, David also painted two of his brilliant history/portrait paintings of the martyrs of the Revolution, Lepelletier de Saint Fargeau and Marat.[41] Political caricatures were a new idiom for the history painter and it is likely that he familiarized himself with caricatures of the period, both French and English (especially Gillray), as well as famous French political prints of the past, such as those of the famous seventeenth-century printmaker, Jacques Callot, whose etchings were well-known.[42] David's stature as the pre-eminent history painter in France helps to explain the finely wrought compositions which are characterized by a delicate drawing technique and the careful construction of a perspectival space in landscape settings. David's engravings, although they use scatological imagery metaphorically, have little else in common with the often crude, simplified, and direct style of many French Revolutionary caricatures, especially the etchings produced in both France and England. David's prints are complex and not easily legible in terms of their meaning at first glance, which explains why David appended rather lengthy captions that identified various figures and the broader meaning. The audience he had in mind could not have been the broad public, including the sans-culottes, that Carnot and his committee wanted to reach with this commission, but a more elite, educated group to which David himself belonged. Nonetheless, his compositions were well received and on May 18, 1794, the *Comité du salut public* decreed the production of the engravings in 1000 copies each, 500 of each to be produced in color. Thus, a lavish amount of money was spent on them, and David himself was also paid handsomely.[43]

In one of the compositions, David depicts a monumental emaciated devil, described in the caption as an "écorché tout vif" ("flayed figure skinned alive"), the embodiment of "The English Government". Combining the sacred with the blasphemic seems at first incongruous, but the appearance of the devil as the political embodiment of evil misrule goes back to the *diaboleries* associated with French medieval and Renaissance fairs and carnivals. The Devil wears the British crown and

[41] Johnson. *Jacques-Louis David*: Art in Metamorphosis, 70–120.

[42] Boime, "Jacques-Louis David," 74–75.

[43] Ibid., and Cuno, "En temps de guerre," 521–542.

the Order of the Garter and leans on a pitchfork decorated with symbols of British royalty. With medusa-like hair made up of coiling snakes, the Devil stares out at the viewer with a horrific but surprised expression on his fiendish face. This is understandable because it turns out that he is suffering from severe indigestion. Sulfurous clouds are expelled from his bottom, formed by the head of George III, as well as lightning strikes which are expelled from the mouth of George III as vomit that contains the word "taxes". The victims of the Devil's severe indigestion are the hapless British citizens, "free Englishmen" ("anglais libres"), some falling beneath this barrage, others attempting to flee but to no avail. The caption reads:

> The government is personified by the figure of a Devil skinned alive, monopolizing commerce and covered with all the royal decorations. The portrait of the king is located at the rear end of the government which vomits on its people a myriad of taxes which overwhelm them. This prerogative is attached to the scepter and the crown.[44]

The King excretes vomit and gases in the form of taxes on the British people. He is not in charge of the diabolic English government that excretes on its people but instead is placed at the exit of the alimentary canal.

David's second satirical engraving, *The Royal Army of Jugs*, is much more complex, with multiple episodes depicted simultaneously in a composition that utilizes the genre of battle scene paintings. In the foreground, the British army—jugs with human heads and skinny legs—are marching behind their jug-leader, George III, who is being led on a leash by a turkey identified in the caption as "Milord Dindon", William Pitt. To the right in the background, the British cannons are nothing more than huge enema syringes (clystères) used in the eighteenth century to give relief from a wide variety of digestive problems (made infamously comic to the popular imagination by the satirizing of doctors

[44]"Ce gouvernement est représenté sous la figure d'un Diable écorché tout vif, accaparant le Commerce et revêtu de toutes les décorations royales, le Portrait du roi se trouve au derrière du Gouvernement lequel vomit sur son peuple une multitude d'impots avec lesquelles il les foudroye. Cette pérogative est attaché au sceptre et à la couronne". English translation by Boime, ibid., 74. Boime found a source for David's devil in Callot's *Temptation of Saint Anthony*, 73–74.

and medicine in the plays of Molière). To the left in the background a group of sans-culottes squat on the top of a crumbling antique arch in the Roman style. They have lowered their pants and are defecating on the avant-garde of the Royal army of jugs, reminiscent of the episode in Swift's *Gulliver's Travels* when the Yahoos climb trees and defecate on Gulliver from above, mentioned earlier.[45] The sans-culottes are using the British jugs as chamber pots. As they are struck by the weaponized feces, the jugs, now functioning as toilets, break apart at which point monsters emerge, revealing what the British army is really made of. The caption describes this scene in the following terms:

> The avant-garde of the Royal army is defeated at the city gate, caused by the colic of several sans-culottes placed at the top of the gate. In its defeat the avant-garde breaks the jugs from which all sorts of venomous beasts emerge which is the spirit that animates them.[46]

David's prints do fulfill the intent of the commission to call attention to the foreign enemies and make them look ridiculous and impotent and thus counteract fears of an English invasion. But the visual language as well as the accompanying caption reveal the complex content of the satire. *The Royal Army of Jugs*, like *The English Government*, is distant from the much less expensive etchings of the period which were intended for the popular masses rather than an educated elite.

David's scatological prints, in their complexity and intellectual content, are forerunners of those produced by one of the prepotent caricaturists of the nineteenth century, Daumier, whose entire oeuvre is characterized by a striking originality. Louis-Philippe's ascendance to the throne shortly after the brief and abortive second French Revolution of July 1830, brought in its wake political anger, instability, and resistance that found an outlet in the scathing caricatures of the early years of the July Monarchy. Many of the satirical images use the processes of consuming and digesting as a means of critiquing the new

[45] Swift, *Gulliver's Travels*, 216.

[46] "L'avant-Garde de la Royal Armée reçoit un échec à la porte de la Ville qui est occasioné par la colique de quelques sans-culottes placés au haut de la Porte. L'avant-Garde dans sa défaite brise les cruches, dont il ne sort que toutes sortes de Bêtes vénimeuses qui est l'esprit qui les anime".

Fig. 5.7 Honoré Daumier, *Gargantua*, December 16, 1831, lithograph. Bibliothèque Nationale de France, Paris

monarchical government. Daumier's satirical portrayal of Louis-Philippe as *Gargantua* (Fig. 5.7), is a pre-eminent and famous example.[47] At the beginning of this article, we looked at the 1791 caricature of Louis XVI as the modern Gargantua, consuming all the agricultural produce of France and drinking the blood of a subject instead of wine. As we have seen, Rabelais' Gargantua, a giant of vast appetites with an enormous ability to ingest and digest massive quantities of food and drink, came to be used in late eighteenth- and nineteenth-century France as a trope for the extremes of excess in the monarchy and ruling classes. Daumier's 1831 caricature, like that from 1791, depicts the glutton with a grotesque body and gaping mouth, bearing the features of the new king, Louis-Philippe. Here the king's face is enormously distorted into the shape of a huge pear. This is a visual pun on *poire*, French slang for simpleton, an image created by Philippon (founder of *La Caricature*,

[47] See Childs, "Big Trouble," 26–37.

known for its scathing political cartoons), and as is well known, Daumier worked for Philippon.[48] Louis-Philippe's body is far more distorted and obese than that of Louis XVI. He is mammoth compared to his Liliputian-type subjects who serve him. And he is not seated at table dining with his family but enthroned on a toilet. His tongue is not an organ of mobility used to push food toward the throat in the digestive process but instead is a monumental rigid ramp that extends from his mouth to the ground—there is no end to the continual consumption enabled by this industrialized organ.

The earlier 1791 Rabelaisian Gargantua, that uses an agrarian metaphor of consumption, is here replaced by the devouring of money. Daumier's Gargantua does not consume food but swallows whole the baskets of coins brought up to his mouth by workers who struggle up the steep ramp/tongue. At the base of the tongue, baskets are filled by ministers who gather the coins from the poor, the starving, and the working class. This is a scathing indictment of the King growing rich and fat from the cruel taxation of the poor. While devouring the taxes of the poor, Louis-Philippe defecates paper awards to government officials who gather them as they are excreted. These include military commissions, the nominations of peers, and appointments to the prefecture (the National Assembly building can be seen in the background behind them). The king consumes all that the poor can feed him and digests what they can offer to give to his government appointees. But his massive size implies that he keeps a great deal of what he consumes in his own body. The metaphor of consuming, digesting, and excreting is thus used to demonstrate the corruption and debasement of the entire government. As Elizabeth Childs notes: "Daumier's gluttonous Louis-Philippe rules not in the elevated spheres of politics, power, and philosophy but in the lower spheres of food, fat, and feces".[49] I would take this a step further for in this image Daumier conflates the Rabelaisian "bodily underworld" of invisible digestive functions with the world of politics in which the rich metaphorically consume the poor. There can be no elevated sphere of politics when France is governed by a monstrous and

[48] Ibid., 33–34. Daumier's lithograph of Gargantua, likely originally intended for *La Caricature*, was not published in the journal but was on exhibit in a print shop in Paris.

[49] Ibid., 28.

insatiable giant whose impoverished subjects serve his boundless need for consumption. It is not surprising that Daumier's scatological image famously led to his arrest and imprisonment for "arousing hatred of the king and his government".[50]

Daumier was daring in his use of the face and body of the king eating and defecating as a critique of the economic and social problems generated by the new King and his regime from the beginning of the July Monarchy. But he was surpassed in his use of scatological imagery by his fellow caricaturist who also worked for Philippon's journal, Charles-Joseph Traviès.[51] In an 1832 print, Traviès represented Louis-Philippe as a pear seated atop a *pot de mélasse*, the crock of molasses (the slang term for excrement) which is placed on a chair.[52] The features of the king inscribed on the pear face the viewer. The French people from differing levels of society gawk at this literal crock of mélasse that embodies the king and his rule.

Another print by Traviès in *La Caricature* from 1832 is entitled *Le Juste Milieu se crotte* (*The Juste Milieu Soils Itself*).[53] Two ministers dressed as Pierrot and Harlequin from the Commedia del' Arte, carry a large crock of fecal matter in the shape of a pear. Louis Phillipe, now pictured as the pot of molasses itself filled with excrement that has been gathered from cesspools, streets, and buildings in Paris, is being carried out of town at night by the vaudeville figures who walk through effluvia leaking from sewage pipes into the street. The minister/waste collectors remove the stinking waste that constitutes Louis-Philippe and his regime under cover of night but fail to evade the gaze of the restless gathered crowds and the urchin who chases them out of town. The print speaks to the anger of the citizens, to the embarrassment of the ministers and to the farce-like nature of the regime. Weisberg has studied this and similar images in the context of the terrible sewage problems in Paris at the time. New laws for removing excrement at night from the city beleaguered by this problem had been enacted and the print would have resonated with the populace at this level as well.[54]

[50] Ibid., 26–27.

[51] Weisberg, "In Deep Shit," 36–40.

[52] Ibid., 37, Fig. 2.

[53] Ibid., 37–38, Fig. 4.

[54] Ibid.

I would like to conclude with one of the most scathing images that used consuming and digesting as political metaphors: Grandville's *Digestion of the budget: Administrative, political, moral and above all, economic work* (*Digestion du budjet [sic]:Travail administratif, politique, moral et surtout économique*) published in *La Caricature* on May 24, 1832.[55] Grandville uses a more naturalistic style here, in significant contradistinction to the grotesque body or exaggerated forms we have seen in caricatures of this type. He depicts Louis Philippe from the back, seated on a toilet, sucking food from a vat into which wealth is being poured that he consumes and digests. The waste he excretes is being siphoned off by his deputies in descending level of importance from left to right. At the far right, the poorest people from society attempt to extract whatever is left through very narrow siphons, almost straws. Part of the excreted wealth is also siphoned by those in charge of secret funds, hidden in an underground vault, that we see in the foreground right. The process of digestion is thus used to describe how the body politic works. This was Grandville's metaphorical use of digestion to show the essence of "trickle-down economics". It provides an acerbic commentary on what was perceived as a regnant economic monism linked to the recrudescence of a gluttonous monarchism that succeeded with signal swiftness the great Revolution of 1789 and the abortive Revolution of 1830.

Digestive metaphors, from bodily excretions such as excrement and vomit to an almost atavistic cannibalism, were powerful agents of satirical attack on those in power and functioned as a means of inspiring, inciting, and empowering the powerless. The prints also functioned didactically to shine light on political corruption and inequities using humorous means. From the late eighteenth to the early nineteenth century in France, as we have seen, the visual and literary tradition of using consuming and digesting as a conduit for cultural and political critique parallel investigations in the medical sciences into the complexities of the digestive process. The images we have been examining reveal the distress of the body politic and sometimes its deliquescence through tropes, metaphors, and analogies of insatiable appetites as well as malfunctioning digestive systems.

[55] Childs, "Big Trouble," 33, Fig. 8.

Bibliography

Arasse, Daniel. *La guillotine et l'imaginaire de la Terreur*. Paris: Flammarion, 1987.

Bakhtin, Mikhail. *Rabelais and His World*. Translated by Helene Iswolsky. Bloomington: Indiana University Press, 1984.

Baridon, Lauren, and Martial Guédron. *L'Art et l'histoire de la caricature*. Paris: Citadelle & Mazenod, 2009.

Boime, Albert. "Jacques-Louis David, Scatalogical Discourse in the French Revolution and the Art of Caricature." In *Politique et Polémique: La Caricature française et la Révolution, 1789–1799*, edited by James Cuno, 69–86. Los Angeles: Grunewald Center for the Graphic Arts, UCLA, 1988.

———. "Jacques-Louis David, Scatalogical Discourse in the French Revolution and the Art of Caricature." *Arts Magazine* 62, no. 6 (1988): 72–81.

Childs, Elizabeth. "Big Trouble: Daumier, Gargantua, and the Censorship of Political Caricature." *Art Journal* 51, no. 1 (Spring, 1992): 26–37.

Cuno, James. *Politique et Polémique: La Caricature française et la Révolution, 1789–1799*. Los Angeles: Grunewald Center for the Graphic Arts, UCLA, 1988.

———. "En temps de guerre: Jacques-Louis David et la caricature officielle." In *David contre David*, vol. I, edited by Régis Michel, 521–542. Paris: Gallimard, 1993.

De Greve, Marcel. *La réception de Rabelais en Europe du XVI au XVIII siècle*. Paris: Champion, 2009.

Diderot, Denis. "Notes on Painting." In *Diderot on Art*, vol. 1, translated by John Goodman, 193. New Haven and London: Yale University Press, 1995.

Diderot, Denis. *Oeuvres completes de Diderot*, vol. 10. Paris: Garnier frères, 1875–1877.

"Digestion." *Dictionnaire des sciences médicales*, vol. IX, 354–451. Paris: Panckoucke, 1814.

Graber, Wilhelm. "Swift's First Voyages to Europe: His Impact on Eighteenth-Century France." In *Reception of Jonathan Swift in Europe*, edited by Hermann Real, 5–16. London and New York: Continuum, 2005.

Halliday, Tony. *The Temperamental Nude: Class, Medicine and Representation in Eighteenth-Century France*. Oxford: Voltaire Foundation, 2010.

Hunt, Lynn. "La psychologie politique des caricatures révolutionnaires." In *Politique et Polémique: La Caricature française et la Révolution, 1789–1799*, edited by James Cuno, 33–42. Los Angeles: Grunewald Center for the Graphic Arts, UCLA, 1988.

Johnson, Dorothy. "Anatomie, réalité, idéalité dans l'art français autour de 1800." In *Ecrire les sciences*, edited by Martial Guédron and Isabelle Laboulais. Special Issue, *Etudes sur le 18e siècle* 42 (2015): 177–192.

———. "The Body Speaks: Anatomical Narratives in French Enlightenment Sculpture." In *Body Narratives. Motion and Emotion in the French Enlightenment*, edited by Susanna Caviglia, 29–62. Turnhout: Brepols Publishing, 2017.

———. "Visceral Visions: Intersections of Art, Pedagogy, and Politics in Revolutionary France." In *Bellies, Bowels and Entrails in the Eighteenth Century*, edited by Rebecca Barr and Sylvie Kleinman, 294–310. Manchester: Manchester University Press, 2018.

Lee, Jae. *Swift's Scatological Satire*. Albuquerque: University of New Mexico Press, 1971.

Reichardt, Rolf, and Hubertus Kohle. *Visualizing the Revolution. Politics and the Pictorial Arts in Eighteenth-Century France*. London: Reaktion Books, 2008.

Stafford, Barbara. *Body Criticism. Imaging the Unseen in Enlightenment Art and Medicine*. Cambridge, MA and London: MIT Press, 1991.

Swift, Jonathan. *Gulliver's Travels*. Oxford and New York: Oxford University Press, 1994.

Vila, Anne. *Enlightenment and Pathology. Sensibility in the Literature and Medicine of Eighteenth-Century France*. Baltimore and London: Johns Hopkins University Press, 1998.

Viola, Paolo. "The Rites of Cannibalism and the French Revolution." In *People and Power: Rights, Citizenship and Violence*, edited by Loretta Valtz Mannucci et al., 157–174. Milan: Dipartimento di storia della società e delle istituzioni, Università degli studi di Milano, 1992.

Vovelle, Michel, ed. *Les Images de la Révolution française*. Paris: Publications de la Sorbonne, 1988.

Weisberg, Gabriel. "In Deep Shit: The Coded Images of Traviès in the July Monarchy." *Art Journal* 52, no. 3 (Autumn 1993): 36–40.

Williams, Elizabeth. *A Cultural History of Medical Vitalism in Enlightenment Montpellier*. Burlington, VT: Ashgate, 2003.

———. "Food and Feeling: Digestive Force and the Nature of Morbidity in Vitalist Medicine." In *Vital Matters: Eighteenth-Century Views of Conception, Life, and Death*, edited by Helen Deutsch and Mary Terrall, 203–221. Toronto: University of Toronto Press, 2000.

———. *The Physical and the Moral: Anthropology, Physiology and Philosophical Medicine in France, 1750–1850*. Cambridge: Cambridge University Press, 1994.

CHAPTER 6

Being "Hangry": Gastrointestinal Health and Emotional Well-Being in the Long Nineteenth Century

Emilie Taylor-Brown

"Shakespeare was not far wrong when he said that an undigested pea might ruin an empire", asserted the author of an article published in the *St James' Magazine* in 1868.[1] The article, which was entitled "Dyspeptic Saints", argued that the "tyranny of the Inquisition" and Mary the First's burning of Protestants might be attributed to improper attention to the "philosophy of the stomach". Bloody Mary, argued the author, injured her constitution by practising fasting, a fact that explains her

[1] "Dyspeptic Saints," 26.

Dr Emilie Taylor-Brown would like to gratefully acknowledge that the research leading to this publication has received funding from the European Research Council (ERC) under the European Union's Seventh Framework Programme (FP7/2007-2013), Grant Agreement Number 340121.

E. Taylor-Brown (✉)
St Anne's College, University of Oxford, Oxford, UK

M. Mathias and A. M. Moore (eds.), *Gut Feeling and Digestive Health in Nineteenth-Century Literature, History and Culture*, Palgrave Studies in Literature, Science and Medicine, https://doi.org/10.1007/978-3-030-01857-3_6

political violence and sealed her fate as the unhappy and childless wife of a Spanish prince. "And who, remembering the fate of Mary, Queen of Scots", they continue, "can say that Henry the Eighth's other daughter, the royal Virgin Elizabeth was of digestion sound?"[2] The pairing of emotional violence and poor dietetic practice is a linkage that we might recognise in the twenty-first concept of "hanger": being angry because you are hungry—or as dictionary.com puts it, "feeling irritable or irrationally angry as a result of being hungry".[3] This portmanteau appeared on urban dictionary as early as 2003 and has since been added to online dictionaries published by Oxford and Cambridge university presses, Macmillan, Collins and Random House.

Despite the contemporary etymology of the word "hanger", hunger and anger are connected experiences that have a long-standing cultural history, from the political violence of the French Revolution—which recalls the famously mistranslated "let them eat cake"—to the devastating protests of the "hungry forties". Peter Gurney connects these two historical moments by recognising the incendiary role of the failure of a compassionate response to working-class hunger. Marie Antoinette's misattributed words and Robert Bullock Marsham's flippant remark in 1842 that, while workers were unable to buy bread they at least "rejoiced in potatoes", voice a stark class divide.[4] This divide, for Gurney, as well as for scholars like John Bohstedt and Charlotte Boyce, is underpinned by the reality of hunger as a culturally and discursively constructed phenomenon that has historically had political power in mobilising the disenfranchised.[5] However, as Bohstedt argues, in order to provoke food riots, hunger must be accompanied by other sociopolitical components, such as the ability to deliver collective action and the confidence that the benefits would outweigh the punishments.[6] Hunger alone does not produce political activism.

[2] "Dyspeptic Saints," 27.

[3] "Hangry," *Dictionary.com*.

[4] Peter J. Gurney, "Rejoicing in Potatoes," 99–136.

[5] Boyce argues that "hunger is both material condition and cultural construct." She explores the difficulties in representing working-class hunger to a middle-class audience in "Representing the "Hungry Forties" in Image and Verse: the Politics of Hunger in Early-Victorian Illustrated Periodicals," 421–449.

[6] See, John Bohshedt, "Food Riots and the Politics of Provisions in World History."

The anger or emotional violence precipitated by hunger is, in these cases then, ideological: a rational and politically inflected response to a larger social problem. However, I want to focus on a more physiologically embodied form of emotional violence made explicit in the "Dyspeptic Saints" article. This pairing of hunger and anger appeared in debates about the interconnectedness of the gastrointestinal and psychological or neurological systems, and so positioned anger in a complex relationship with digestion and appetite. In this essay, I explore the dynamic relationship between digestion and emotional well-being in a variety of nineteenth-century British texts, arguing that the lexis of anger was one of the many lenses through which writers strove to pin down the mechanics of digestive health.

The author of "Dyspeptic Saints" noted that "there was want of due balance betwixt stomach and brain in the case of the great Protestant Queen Bess",[7] a dynamic understanding of selfhood that values parity. This played into a rhetoric of balance that underscored much dietetic discourse throughout the century, not to mention wider discussions of medical and social selfhood. As Fay Bound Alberti notes, until at least the early nineteenth century, the humoral model of health dominated medical discourse. This model relied on the notion of humoral balance—a product of both innate characteristics and environmental factors such as air, diet, exercise and sleep.[8] Throughout the nineteenth century, this notion of bodily balance was continuously recast, at first in response to notions of organic sympathy borrowed from eighteenth-century neurology, and later in response to developments in gastric biology. George Cheyne had famously explicated what he termed "the English Malady" in 1733, a "class and set" of nervous disorders that he associated with "the richness and heaviness of our food", "inactivity and sedentary occupations" and "the humour of living in great and populous, and consequently unhealthy towns".[9] For Cheyne, "dietetic management" was of paramount importance in curing nervous illness, and he lamented the disrepute into which medical dietetics had fallen.[10]

[7] "Dyspeptic Saints," 27.

[8] Fay Bound Alberti, *This Mortal Coil*, esp: 10–13.

[9] George Cheyne, *The English Malady*, i–ii.

[10] Cheyne, *The English Malady*, 149–156.

Following Cheyne's influential thesis, interest in "the English malady" intensified and it became, as Glen Colburn has argued, "a malleable cultural metaphor that could be adapted to a wide range of social [and medical] diagnoses" being used almost interchangeably with hysteria, hypochondria, and later neurasthenia and nervous dyspepsia, and codifying a range of national and political concerns.[11] Cheyne's diagnostic framework in which digestion featured as a central cause of nervous illnesses was elaborated by numerous physicians in the following century. In the early nineteenth century, English surgeon John Abernethy published his *Surgical Observations* in which he championed the theory of gastric sympathy—that local nervous irritation could derange the digestive organs, which would in turn disorder the wider constitution of an individual.[12] Scottish physiologist, Alexander Philip, also endorsed the power of "gastric sympathy", arguing that indigestion might assume an "inflammatory character" and transform from "a mere nervous affection" into "the source of all mischief". He considered indigestion to be a disease "not of any one set of organs but of the whole system".[13]

This position echoed that of French materialist, François Broussais, who believed that the prevalence of gastric symptoms in pathology could be explained by a system in which inflammation of the gastrointestinal tract was passed sympathetically to other organs. In his *Principles of Physiological Medicine* (1832), he claimed that "he who does not know how to manage irritability of the stomach, will not know how to treat any disease. A knowledge of gastritis and of gastro-enteritis is therefore key to pathology".[14] In these medical discourses, the stomach was placed within a sympathetic network of organs, encouraging a view of the body as being composed of discrete but reactive components.

Broussais also noted that "when the stomach is not sufficiently stimulated by food [hunger occurs, causing] an irritation in this organ, in which the brain partakes [...] such are the fury and mental exaltation of starving persons".[15] Clearly, good dietetic practice is a fundamental tenet of bodily *and* mental health for Broussais. The centrality of the

[11] Glen Colburn, "Introduction," in *The English Malady: Enabling and Disabling Fictions*, 2.

[12] John Abernethy, *Surgical Observations.*

[13] Alexander Philip, *On the Treatment of a More Protracted Case of Indigestion*, 3–6.

[14] F. J. V. Broussais, *Principles of Physiological Medicine*, 525.

[15] Broussais, *Principles of Physiological Medicine*, 524.

gastrointestinal region and its role in precipitating emotional and physical trauma reverberated in popular culture and commercial medicine. In 1827, *The Electic Review* asserted that "so general a complaint is Indigestion in this country and so much does it influence other diseases, that there could not perhaps be a more useful treatise than one on the manner in which other diseases are influenced by their concurrence with it",[16] while an author for *The Critic* "tearfully acknowledge[d]" in 1859 that indigestion was "the most common disease met with" in Britain.[17] An advert for Revalenta's Arabica Food in 1849 charged dyspepsia with, among other things, causing "palpitation of the heart, inflammation and cancer of the stomach, nervous headaches, deafness, noises in the head and ears, pains in almost every part of the body", "low spirits", "general debility", "dislike to society", "exhaustion", "melancholy", suicidal thoughts and even insanity.[18] All of this could be solved, the advert asserted, by "a permanent restoration of the digestive functions".

Digestion as a process had received renewed attention following the British publication of William Beaumont's *Experiments and Observations of the Gastric Juice and the Physiology of Digestion* in 1838, which highlighted the damaging effects of, among other things, "strong mental emotions, such as anger" to the digestive capacity, further demonstrating the interconnectedness of emotional and physical health.[19] Beaumont attributed the presence of bile in the stomach to the effect of anger following "violent passionate emotion" (149). His text was groundbreaking because it revealed, for the first time, the inner workings of digestion in a live body. It catalogued a series of experiments that Beaumont, an American army surgeon, had carried out on a Canadian man named Alexis St Martin, who had been shot in the stomach with a musket. The wound never fully healed over, leaving a hole through which the digestive processes could be witnessed in real time, and in an otherwise healthy body.

In his notes to the edition, Scottish physician Andrew Combe argued that it was "useful to notice the analogy which subsists between the

[16] "On the Treatment of the More Protracted Cases of Indigestion," 405.

[17] "A Physician's Notebook," 343.

[18] [Advertisement] *The Athenaeum*, 343.

[19] William Beaumont, *Experiments and Observations on the Gastric Juice*, 315.

process of digestion and that of purely chemical solution".[20] This analogy formed the basis of "gastric chemistry", an amorphous field of study that concerned itself with the chemical makeup and digestibility of food. From analytic chemists involved in marketing dietetic bread,[21] to exposés revealing the extent of nineteenth-century food adulteration, food was increasingly submitted to the legitimising gaze of laboratory science. In 1829, *The London Magazine* had published an article, which expressed concern that the growing practice of subjecting food to the tools and methodologies of the laboratory threatened to make a once familiar constituent of life "furiously anathematized".[22] Beaumont's publication catalysed this trend, which had been gaining traction following exposés of food adulteration like Frederick Accum's *A Treatise on Adulterations of Food and Culinary Poisons* (1820).[23] This increasingly scientific treatment of food and digestion,[24] along with: an expansion of local and national food regulation[25]; developing ideas about vegetarianism[26]; changes in understandings of gastric biology; and the pathologisation of corpulence throughout the nineteenth century[27]; contributed to the prominence of gastrointestinal health in the popular imagination. Discussions of the stomach and its somatic and psychological influence were housed in medical textbooks and science review pieces, but also in novels and poems, in periodical fiction and in general

[20]Beaumont, *Experiments and Observations on the Gastric Juice*, 303.

[21]See, for example, [Classified Ad 29] *Manchester Guardian* (2 October 1897), 4.

[22]John Scott and John Taylor, eds., "Every Man's Master," 347.

[23]Also later: John Mitchell's *A Treatise on the Falsification of Food* (1848), and Arthur Hill Hassall's *Food and Its Adulterations* (1855) and *Adulterations Detected* (1857), which exposed the extent of the adulteration of food to the general reader. Following a year-long anti-adulteration campaign, "The Sale of Drugs and Food Act" was passed in 1875 and further amended in 1899.

[24]See also, Justus von Liebig, *Animal Chemistry*; and Mark Finlay's historical overview: "Quackery and Cookery: Justus von Liebig's Extract of Meat and the Theory of Nutrition in the Victorian Age."

[25]See, for example, Michael French and Jim Phillips, eds., *Cheated Not Poisoned?: Food Regulation in the United Kingdom, 1875–1938.*

[26]See, for example, James Gregory, *Of Victorians and Vegetarians: The Vegetarian Movement in Nineteenth-Century Britain.*

[27]David J. Hutson explores the ways in which body weight became a central concern in medical and popular spheres in "Plump or Corpulent? Lean or Gaunt? Historical Categories of Bodily Health in Nineteenth-Century Thought."

interest essays, in political satire, and in commercial advertisements. Consequently, the mechanics of digestive health found visibility in the crosstalk between disciplines, with generic conventions offering new discursive relationships.

Sydney Whiting's mid-century fictional autobiography, *Memoirs of a Stomach* (1853), gave voice to the digestive organ in a bid to educate the public about the importance of gastric health and its interrelations with other bodily processes. As Joyce Huff notes, the text initially appears to "disrupt dominant models of Cartesian dualism"[28] by elevating the stomach to a position of seemingly autonomous power: Mr Stomach considers himself to "hold a superior position to [his] help-mate, Mr Brain".[29] However, Whiting complicates this reading by codifying multiple forms of agency into the narrative and by using the metaphor of a "double set of electric wires"[30] to explicate the connection between the stomach and the brain—an invocation of telegraphy that was common across genres. Physician Edward Johnson, in his 1849 treatise on hydropathy, had similarly referred to this gut-brain connection using the communicative technology: "Now the brain is London, and the stomach is Bristol, and the gastric portion of the pneuma-gastric nerve is the [telegraph] wire".[31]

The use of such metaphors positioned gastrointestinal health in relation to modernity, making the nervous body synonymous with a fully connected modern society. Laura Salisbury and Andrew Shail have discussed the relationship between neurology and modernity as one of "mutual constitution", in which the neurological self and the "modern" self were "cogenerative".[32] Such constructions causatively linked emotional, physical and national well-being, while at the same time labelling "nervous" diseases characteristic of "modern" life. By nervous diseases, I refer to diseases that were underpinned by a supposed disruption in the nervous system. While we might now locate gastrointestinal health primarily in the digestive and alimentary organs, in the nineteenth century this was understood, as we have seen, in much broader terms. The nerves connecting these organs and the brain were also brought within the

[28] Joyce L. Huff, "The Narrating Stomach," 75–93.

[29] Sydney Whiting, *Memoirs of a Stomach*, vii.

[30] Whiting, *Memoirs of a Stomach*, 21.

[31] Edward Johnson, *Results of Hydropathy*, 102.

[32] Laura Salisbury and Andrew Shail, "Introduction," 1.

purview of gastrointestinal health by virtue of their role in maintaining the relationships between digestion and emotional well-being. Indeed, as Hisao Ishizuka has argued, gastrointestinal disorder was a vital component of the "nervous experience of the modern era".[33]

The gastric component to nervousness was underpinned by a functional connection between digesting and feeling, an age-old idea that was discussed in an 1867 article in Charles Dickens's periodical *All the Year Round*. The article argued that the seat of the affections was not in fact the heart, but the stomach, since "benevolent feelings towards all mankind are notoriously promoted by a good dinner, [while] a damaged stomach occasions melancholy, disgust, envy, hatred, and all uncharitableness".[34] Recognising its long cultural heritage, the author suggested that one need only observe human nature to divine this fact, noting however that they could bring no "anatomical" proof for it.[35] Six years earlier the periodical had featured the serialisation of Dickens's *Great Expectations*, which had likewise explored the relationship between eating and feeling. We meet Miss Havisham for the second time, after all, amidst a decaying banquet of food that matches her sense of abandonment, and Joe uses food as a means of expressing something like fatherly affection to Pip:

> He always aided and comforted me when he could, in some way of his own, and he always did so at dinner-time by giving me gravy, if there were any. There being plenty of gravy to-day, Joe spooned into my plate, at this point, about half a pint.[36]

More poignantly, Miss Havisham equates her grief with being eaten when she remarks of the feast, "the mice have gnawed at it, and sharper teeth than teeth of mice have gnawed at me".[37]

Kerrie Schoffer and John O'Sullivan argue that Dickens's fiction reveals a perceptive attentiveness to illness, particularly to nervous disorders, an assertion apparently supported by the *British Medical Journal's* obituary in 1870, which described Dickens's "rare fidelity" and the gain

[33] Hisao Ishizuka, "Carlyle's Nervous Dyspepsia," 82.

[34] Charles Dickens, ed., "Stomach and Heart," 439.

[35] Dickens, "Stomach and Heart," 438.

[36] Dickens, *Great Expectations*, 45

[37] Dickens, *Great Expectations*, 111.

it would have been to medicine had he used his "keen[ness] to observe" for "the medical art".[38] It is undoubtedly this keenness of perception that underpins Dickens's portrayal of Camilla in *Great Expectations*, a character who is eager to play the victim and who invokes her husband Raymond as witness to the "nervous jerkings" that she has been suffering from following Miss Havisham's misfortune. Dickens satirises Camilla's affected weakness of constitution by having her exclaim melodramatically "I cried about it from breakfast till dinner. I injured my digestion",[39] and later, "If I could be less affectionate and sensitive, I should have a better digestion and an iron set of nerves".[40] Despite his intended ridicule, Dickens is echoing contemporary dietetic and biological treatises, which warned that "strong mental excitement [...] will temporarily exhaust the digestive powers".[41]

Such assertions were in dialogue with changing understandings of the body, particularly ideas about a finite store of nervous force that supplied the body with energy. As Roy Porter has noted, the Victorian model of supply and demand in relation to nervous illness borrowed its conceptual and linguistic frameworks from contemporaneous developments in energy physics and from dominant economic metaphors.[42] These frameworks encouraged individuals to see their bodies in terms of "resources". Consequently, healthy digestion was often framed as a negotiation with other organic processes, particularly cognitive ones. A faulty digestion might starve the intellect of adequate resources of nervous energy, or vice versa, leading to nervous collapse. An advertisement for Kutnow's powder outlined this problem in 1898 and placed it firmly in relation to what were considered by many commentators to be the conditions of "modern life":

> The necessity of rapid action and abnormal exertion of the nervous system has increased to such a degree that our supply of nervous energy is not equal to demand. This appears to be owing to our failure to confine our diet to that which is easy of digestion and assimilation [...] to over-work

[38] Kerrie L. Schoffer and John D. O'Sullivan, "Charles Dickens: The Man, Medicine, and Movement Disorders," 630.

[39] Dickens, *Great Expectations*, 104.

[40] Dickens, *Great Expectations*, 109.

[41] John Timbs, *Hints for the Table*, 7. See also: Beaumont, 315.

[42] Roy Porter, "Nervousness, Eighteenth and Nineteenth Century Style," 31–50.

> and over-worry and to the unhealthy and artificial conditions under which we live.[43]

The role of diet in supplying nervous energy and the dangers of "forcing" the brains of the overworked and poorly fed were also depicted in Margaret Harkness's 1890 social problem novel *In Darkness London* in which the East End doctor asserts, "why, it is brutal to make children go to school without a proper breakfast, to force the brains of boys and girls who have empty stomachs". It is only due to the "generosity" of the poor to each other that we avoid, the doctor contends, "a generation of idiots".[44] Sally Shuttleworth has written about the dangers of "brain-forcing", particularly in relation to the education of children. She recounts James Crichton Browne's concerns that there had been "a huge rise in nervous disorders and brain disease due to overpressure, which was exacerbated for the poor by ill-nourishment".[45] Despite the Chief Inspector of Schools, J. G. Fitch's, damning refutation, Shuttleworth highlights the "huge social and cultural impact" that Crichton Browne's work had, including a flurry of discussion in the medical and public presses.[46] Harkness's novel upheld Crichton Browne's warning by referring to the experience of East End poverty as one of "adulterated and indigestible food", accompanied by "exciting changes in mental condition [...] violent vibrations between hope and fear".[47] The physical and emotional trauma of these dietary and emotional circumstances was responsible, suggested Harkness, for the wretchedness and brutalism of the poor.

Scenes like these in novels worked to broaden understandings of the labouring classes' experiences of hunger and to draw attention to the complex interplay between the physical and emotional needs of the populace. However, such depictions were not confined to the poor. The fictionalised middle class also experienced embodied psycho-gastric distress and exhibited emotional and physical violence, not in relation to hunger but to more chronic digestive disorders. This is such a common occurrence that it was even a subject of literary complaint. In a review for

[43] "Kutnow's Powder," 7.

[44] Margaret Harkness, *In Darkest London*, 66.

[45] Sally Shuttleworth, *The Mind of the Child*, 135.

[46] Shuttleworth, *The Mind of the Child*, 135–136.

[47] Harkness, 67.

Arthur Marchmont's novel *Madeline Power* (1891), the reviewer laments such a convention: "the hero's father is a man of violent temper who has the gout. In conventional fiction, it is impossible for an old man to have a violent temper without having the gout also. The gout accounts for the temper".[48] In these contexts, anger is pathologised as a symptom of disease, rather than a bona fide emotion.

In *Gout: The Patrician Malady*, Roy Porter and George Rousseau call gout "the most male of all masculine medical maladies" and recognise its literary construction as a "virtual insignia of a ruling patriarchy", particularly in the eighteenth and nineteenth centuries.[49] However, this medical gendering was more complex when it came to broader representations of digestive complaint. Dyspepsia, for example, was a complaint that resisted gender stereotypes. As Gail Houston points out, the Victorian gender-based code of consumption mandated that women should not hunger after food (a proxy for other political and sexual activities) in the same way that men did.[50] Consequently, one might be tempted to label dyspepsia—a condition often associated with excessive consumption—as a male complaint. However, digestive imbalance (with its attendant emotional baggage) was often attributed to men and women in equal measure.

In 1868, *Sharpe's London Magazine* published a poem entitled "Molly's Dream", in which a young woman is plagued by a violent goblin. The poet calls him the "imp of indigestion" and uses a pun on "goblin" and "gobbling" to warn against the perils of overindulgence.[51] Supernatural creatures like this one were commonly invoked to articulate the violence that writers associated with alimentary distress. In this way, disruptions to digestion were understood in physically and emotionally traumatic terms. The goblin in this poem gives imaginative form to Andrew Combe's assertion that in cases of digestive illness, the stomach is "punishing us with the pangs of dyspepsia as a warning to more reasonable conduct".[52] A poem published in *Judy* in 1883 made this idea even more literal with the lines: "A demon stands behind my chair,/And watches what I swallow [...] I taste a tart, the demon digs/me in the

[48] "One Volume Fiction," 413.

[49] Roy Porter and G. S. Rousseau, *Gout: The Patrician Malady*, 143.

[50] Gail Turley Houston, *Consuming Fictions*, xii.

[51] S. C. Hall, ed., "Molly's Dream," 91.

[52] Andrew Coombe, *The Physiology of Digestion*, 243.

chest directly,/As if to say, 'Come stop your rigs/And eat more circumspectly!'"[53] While Combe placed part of the blame for indigestion on overindulgence, as his comment suggests, he also noted the importance of the healthy functioning of the relationship between stomach and brain in guiding our ability to make informed decisions. A disruption in this relationship, such as might be caused by "strong mental emotions and earnest intellectual occupation" (16), could cause an individual to lose the ability to "distinguish [hunger's] true dictates" (24).

Combe associated such disruptions to the gut-brain relationship with the working classes, who "suffer physically, morally, and intellectually from being over-worked and under-fed" (246) and also with the "brain-worn classes" who suffer from "excessive and deleterious cerebral irritation" (265). Such cerebral irritation might upset the appetite, which he held to be the essential mechanism by which animals replenish energy and resources because it "enforce[s] attention to the wants of the system" (10). In *A System of Phrenology* (1834), his brother, George Combe, associated the appetite with the "organ of alimentiveness", which he located in the zygomatic fossa of the brain. "The stomach is to this organ what the eye is to the sense of seeing", he argued, "cut off the communication between it and the brain and the appetite will be lost".[54] He argued that "hunger and desire to eat always go together, just as [...] anger is universally accompanied by the propensity to inflict suffering".[55]

Such a parallel was made literal by his assertion that the "organ of alimentiveness" was a continuation of the "organ of destructiveness" situated just above and behind it. This mapped hunger and violence (the boundaries of which, as Combe noted, were in contention) in topographic proximity.[56] In the appendix to the 1851 edition, Combe used cerebral localisation to explain occurrences of "enormous eating". He relates a case in which a man lured seven strangers into the woods, brutally murdered, and then ate them. Upon dissection, the man was said to have had an abnormally large "organ of alimentiveness". This suggested that the extreme violence of murder-followed-by-cannibalism (a starkly

[53] "The Demon of Dyspepsia," *Judy*, 307.

[54] George Coombe quoted in Jonathan Pereira, *A Treatise on Food and Diet*, "Appendix (A) pg 10," 265.

[55] George Coombe, *A System of Phrenology*, 288.

[56] Combe, *A System of Phrenology*, 176.

transgressive act) was the product of a pathological or non-normative appetite.

In 1876, a writer for Manchester-based journal *The City Jackdaw* also associated violence with the act of eating; however, unlike Combe, who located this pathology in the cerebral tissue, the anonymous author depicted violence as a direct product of gastric derangement. Suffering from a painful dyspeptic episode, he asserted that if he "should ever murder [his] wife and family [...] the fault may be set down to indigestion".[57] He went on to warn readers, "do not, my unknown friend, touch that beefsteak pudding, which has been served to you; there is murder in it [...] there is hatred in that crust, revenge in those lumps of fat, and murder, or at least manslaughter, in those lumps of meat and gristle".[58] While this statement might equally invoke adulteration scandals or early animal rights movements, it is here firmly embedded in the discourses of dietetics. Indigestible food causes what the author identifies as not only digestive disorder, but embodied emotional distress and violent behaviour.

Like many others before him, the author used demonic possession as a proxy for his symptoms. In doing this, he drew on a network of metaphors and analogies that together formed a cultural referent known as "the demon of dyspepsia". The demon derived part of its imaginative aesthetic from cartoons like George Cruikshank's famous 1835 cartoon entitled "indigestion", which featured a man beset by goblin-like creatures. In 1888, it gave its name to Adolphus Bridger's medical textbook on indigestion, and as a cultural figure it bridged discussions in literature, medicine and popular culture because it emblematised both the physical pain and emotional trauma that those suffering from digestive complaint experienced.

As an aesthetic framework it proved popular in part because it neatly reframed folkloric explanations of emotional and physical distress. It was able to reconcile modern medical rhetoric with descriptions, as an author in *Chambers's Edinburgh Journal* noted, of ancient demonic possession. "Science has robbed us of the horns, the hooves, and the tail, but it has, with all its poetry-spoiling discoveries, still left us with the essential demon. The monster is called by nosologists 'dyspepsia' and by the rest

[57] A Hypochondriac, "What to Eat, Drink and Avoid," 379.

[58] Ibid.

of the world, indigestion".[59] Significantly, the writer identifies the possession of "a morose unamiable disposition" and "snappish[ness]"—in other words, poorly regulated and violent emotions—with the presence of chronic indigestion. Emotions and behaviours that had once been attributed to satanic possession were now being viewed as a product of bodily disharmony engendered by digestive distress.

Countless advertisements for curing indigestion and strengthening the stomach also made use of this supernatural analogy. Mother Seigel's Syrup, for example, referred to the first symptoms of chronic indigestion as "the touch of the demon's chilling fingers" and warned readers to "keep out of his clutches".[60] Other writers went further, offering the violent nightmares of indigestion as explanation for a historical belief in the spiritual world. Dickens's *A Christmas Carol* is proffered as an example of the "horrors which imagination can body forth under temporary derangements of the sensory apparatus".[61] The author reads Scrooge's encounter with the ghost of Marley as nothing more than a dyspeptic episode, a "hobgoblin" precipitated by "the imminent festivities of Christmas".[62] Others too read this framework in Dickens; in a synopsis of his short story *The Chimes* (1844), the authors of *Hood's Magazine and Comic Miscellany* note "for this offence [Trotty] is stunned and lectured by the Bells and mobbed by their goblins; and like Scrooge undergoes an awful vision [...] the Porter awakes, and discovers that he has only been dreaming a bad dream, induced by a too hearty dinner of tripe".[63] In the original story, Meg reprimands her father in kind: "don't eat tripe again without asking some doctor whether it's likely to agree with you; for how you have been carrying on".[64] With such literary episodes, it is rather unsurprising that dyspepsia finds its symbol in demon apparitions that often functioned to inspire dietetic (and moral) contrition in their victims. In 1878, *Fun* published a poem entitled "Indigestion! A Christmas Carol", which played on these same connections, using the demon figure to emphasise the violence of digestive complaint:

[59] "Dyspepsia," *Chambers's Edinburgh Journal*, 12.

[60] Mother Seigel's Syrup [Advertisement], 2.

[61] "The Literature of Spiritualism," 1518.

[62] Ibid.

[63] "The Chimes: A Goblin Story," 77.

[64] Charles Dickens, *The Chimes: A Goblin Story*, 79.

Then down on your bosom, with pendulum stroke,
The pudding comes clattering—thud!
The pleasant suspicious your *sternum* is broke
Conduces to curdle your blood,
And there on the top is the demon himself,
A goodly king for it all!
You clutch at the clothes to get rid of the elf—
That counterpane is your pall!
    And you toss and rear
    First there, then here.
And squirming, and foaming, and groaning you rave;
    But all in vain,
    You'll *keep* the pain;
The demon your agonised efforts will brave.[65]

In this portrayal of the demon of dyspepsia, the poet draws on the association between dreams and indigestion, gesturing to the chest-sitting incubus famously depicted in Henry Fuseli's *The Nightmare* (1781). As Jennifer Ford has argued, the connection between physiological processes and the imagination was the subject of lively debate in the late eighteenth and early nineteenth centuries.[66] Dreams had long been associated with bodily processes like digestion and artistic depictions of dreams in the late eighteenth century, such as Fuseli's painting, aimed to depict not the symbolic content of dreams, but the physical experience of them.[67] Sally Shuttleworth notes how, from the mid-nineteenth century onwards, there was a shift in psychiatry away from digestive explanations for mental phenomena like bad dreams. As she points out, while British physician Charles West attributed night terrors to digestive dysfunction in 1848, just six years later in 1854 he had "radically reframed" his understanding of night terrors in the light of a new interest in mental disorders of childhood.[68] However, it is within the earlier tradition of connecting gastric and psychological processes that the "demon of dyspepsia" operates. In popular discussions of gastrointestinal health, dreams were assigned the dual task of articulating—by analogy—the felt experience of dyspeptics, as well as codifying the actual hallucinatory

[65] Henry Sampson, ed., "Indigestion! A Christmas Carol," 252. Emphasis in original.

[66] See, Jennifer Ford, *Coleridge on Dreaming*.

[67] See, Nicola Bown, "What Is the Stuff That Dreams Are Made Of?," 152–153.

[68] Sally Shuttleworth, *The Child of the Child*, 42.

component that accompanied some forms of the disorder. In this capacity, the discourses surrounding digestive distress formed a part of modified understandings of hypochondriasis.

In "Hypochondria: Medical Condition, Creative Malady", literary historian, Stephen Heath, gives a useful overview of the history of the disease, which was recognised by the Greeks as a pathology linked to and deriving its name from an anatomical space within the abdomen called the hypochondrium. Famously associated with melancholia and fear of death, it was also characterised by gastrointestinal symptoms. Heath charts its formulations from the sixteenth to the nineteenth century and beyond, as it became increasingly psychologised, and highlights its overlap with neurasthenia, hysteria and psychosis, to paint a picture of what he considers today to be a "culturally extensive anxiety" about modern life.[69] This anxiety about the damaging impact of modernity is, of course, not unique to the twenty-first century. Since George Cheyne's *The English Malady* in the eighteenth century, which had associated nervous conditions like hypochondria with "universal trade" and urban expansion,[70] medical and non-medical commentators had increasingly associated disease with what they perceived to be the conditions of modernity or civilisation. By the mid-nineteenth century, a version of hypochondriasis that privileged the role of digestive disorder had risen in visibility with the name "dyspeptic hypochondriasis", or sometimes "hypochondriacal dyspepsia". Although gastrointestinal symptoms had always been a part of the condition, this heightened focus on them suggested that poor dietetic practice stood at the bottom of what was increasingly characterised as a psychosomatic disorder.[71]

Scottish physician Anthony Todd Thomson had recounted the experiences of the hypochondriac in his "Extracts from the Notebook of a Physician" published in the *New Monthly Magazine* in the 1830s. In communicating the patient's embodied experience, he used part of the imaginative aesthetic that was in this same period being employed in relation to dyspepsia. In both cases, somatic and psychic trauma was expressed as a violence perpetrated by demonic figures:

[69] Stephen Heath, "Hypochondria: Medical Condition, Creative Malady," 919.

[70] Cheyne, *The English Malady*, ii.

[71] George Rousseau has written on the connection between hypochondriasis and digestive complaint with reference to Samuel Taylor Coleridge in "Coleridge's Dreaming Gut: Digestion, Genius, Hypochondria."

> Like the gnomes of the Rape of the Lock, each has his peculiar province. One pierces with his forked tail, the supra-orbital nerve [...] another whips her shoulder [...] a third digs into a joint [...] a fourth has the temerity to descend into the stomach [...] whilst numbers turn the brain upside down; or reckless, pinch the nerves, exulting in the tremblings, pantings, vertigoes, palpitations, burning heats, and intolerable alarms, which their mischievous pranks awaken in the mind, and the bodily frame of the fair invalid [...] she [the patient] really begins to think that her head must be a steam-engine; that little boys are setting off balloons in her stomach; she fears she will become insane.[72]

Thomson's reference to the steam engine here betrays the disease's associations with the Industrial Revolution, which was characterised by the increasing use of steam power. Indeed, much later in the century, American neurologist George Miller Beard would trace nervous conditions to "modern civilisation" which he considered to be synonymous with "steam-power, the periodical press, the telegraph, the sciences, and the mental activity of women".[73] Thomson elsewhere refers to his patient's symptoms as a "host of blue devils", a striking invocation of Cruikshank's depiction of indigestion. He further points to the shared root of hypochondriasis and dyspepsia when he notes that the sufferings of the hypochondriac are due to "physical disturbances of the bodily system [...] the same which generate dyspepsia, and diseases of the alimentary canal".[74] He asserts that although the supernatural analogies put forward by the hypochondriac (and clearly here perpetuated by the medical profession) appear "ludicrous", they are "nevertheless the recital of *real* morbid feelings". In other words, the demonic figures that haunt the imaginative aesthetic of digestive ill-health corresponded to a real felt experience.

The current definition of hypochondriasis as set out in the diagnostic and statistical manual of mental disorders (DSM-IV-TR) is enlightening here. Included among the somatoform disorders, it is defined as a "preoccupation with fears of having, or the idea that one has, a serious disease based on misinterpretation of one or more bodily symptoms".[75]

[72] Anthony Todd Thomson, "Extracts from the Note-Book of a Physician—No. I," 195.

[73] George Miller Beard, *American Nervousness*, vi.

[74] Thomson, *Extracts*, 196.

[75] Michael B. First and Allan Tasman, *DSM-IV-TR Mental Disorders: Diagnosis, Eitology, and Treatment*, 1002.

This misinterpretation was, in the nineteenth century, often posited as a product of the fraught relationship between stomach and brain. An article in *Bow Bells* in 1871 argued that "mental anxiety and pecuniary embarrassments" could lead to indigestion, while physician John James asserted that dyspeptics are susceptible to the "keenest sensations of an unpleasant nature. They are querulous, irritable, irascible, suspicious and desponding".[76] These "hypochondriac affections", he argues, manifest when disease has taken seat in the "auxiliary digestive organs".[77]

Beard would later use dyspepsia as an analogue for insanity, both of which he associated with "civilised lands", "over-work" and "over-worry".[78] Thomson had also made this connection, arguing that his patients often suffered "from illusions similar to those which occur in insanity; that they see individuals and things, which have no real existence".[79] He writes of a Scottish lawyer who imagined that he was always attended by a little gentleman in court dress, which over time transformed into a haunting skeleton. He also writes of a lady who would see a black dog sitting on the table near her—both phantoms of a "diseased imagination". Such phenomena might remind us of Joseph Sheridan Le Fanu's short story, "Green Tea" (1869), in which the protagonist is dogged by the apparition of a demon monkey. He tries to rationalise it as a "symptom of nervous dyspepsia", having read about it in the medical journals.[80] Indeed, Thomson's notebook provides the basis for just this plot. He notes that hypochondriasis (also known as nervous dyspepsia) has two causes, physical and moral:

> The physical causes are the same which generate dyspepsia, and diseases of the alimentary canal, originating chiefly in sedentary occupations, and an improper use, or rather an abuse, of tea, coffee, and similar diluents. Literary men, and those studiously inclined amongst the clergy, indulge in the use of the herb of China to an extraordinary degree; the consequences of which [...] invariably terminates in hypochondriasis.[81]

[76] John James, "ART I. An Essay on Indigestion," 4.

[77] James, 5.

[78] George Miller Beard, "The Extent of Insanity," 668–699.

[79] Thomson, "Extracts from the Note-Book of a Physician—No. I," 195.

[80] J. Sheridan Le Fanu, "Green Tea," 58.

[81] Thomson, "Extracts from the Note-Book of a Physician—No. I," 197.

The protagonist's condition in "Green Tea" then corresponds to Thomson's understanding of dyspeptic hypochondriasis. Jennings is a clergyman who had been studying ancient metaphysics and drinking green tea—a recipe for hypochondriasis. The "diseased imaginations" of Thomson's patients and Le Fanu's protagonist blur the boundaries between dyspepsia, hypochondriasis, and insanity—an epistemological uncertainty that drives the plot of "Green Tea".

Irish physician, Richard Madden, had too erected this framework, asserting that these conditions were part of a sliding scale of ill-health: "hypochondria is the middle state between the vapours of dyspepsia and delusions of monomania".[82] Le Fanu's characterisation of the demon monkey as persistently violent and malicious (as well as supernatural) again drew upon the aesthetics of the demon of dyspepsia and thus further encoded discourses of gastrointestinal health into a narrative traditionally read for its depictions of delusional insanity. This entanglement was consistent with many physicians' assertions that hypochondriasis was in fact "an aggravated form of dyspepsia,"[83] and that hallucinations belonged just as rightly to the symptomology of digestive disorder as they did to that of psychological illness. In their review of Madden's book, the *Quarterly Review* criticised Madden for overstating the influence of dyspepsia on the mind and body in his suggestion that "giddiness" and "petulance"—"languor" and "irascibility"—"dejection" and "revenge"—"headache" and "artifice"—"palpitation of the heart" and "parsimony of paper" were symptoms of the condition.[84]

A 2010 paper on the psychopathology of functional gastrointestinal disorders highlights the importance of late nineteenth-century psychology to current theories of gastroenterological "psychosomatics". In the 1880s, American and Danish psychologists William James and Carl Lange independently forwarded theories of emotion that implicated physiological processes directly in the experience of emotion. For them, the physical changes of the body in response to emotional stimuli were not the by-product of emotion but rather were constituent of the

[82] Richard R. Madden, *The Infirmities of Genius*, 153.

[83] Madden, *The Infirmities of Genius*, 155.

[84] "ART. II. – The Infirmities of Genius Illustrated by Referring the Anomalies of the Literary Character to the Habits and Constitutional Peculiarities of Men of Genius," 42.

feeling of that emotion.[85] In their paper of 2010, Lukas van Oudenhove and colleagues note that these theories critically arose out of an international medical community engaged in constant debate about the direction of influence between GI disturbances and psychopathology. Read in this context, William Beaumont's assertion that "violent passions" such as anger cause secretion of bile, suppress the secretion of gastric juice, irritate the stomach lining and interrupt the process of digestion, intervenes not just in the physiology of digestion, but in the physiology of emotions and their mutual intertwinement. Not only do these discourses bring emotions like anger into the lexis of the pathological, but they also interrogate and attempt to modify the boundaries between feeling and emotion.

Gastrointestinal ill-health thus becomes a much richer and more nebulous category of complaint, its difficulty of expression reflected in popular culture by the anthropomorphising of gastric symptoms as supernatural demons and the use of violence to codify psychosomatic distress and alimentary guilt. In 1846, *The London Journal* published a poem sent in by a reader that recounted his experience of being hungry on his sickbed. "Why dread the angry cloud/Of thunder, tempest, rain/When there's an element as loud/That rages in our brain?" he wrote, lines that recognised the emotional distress of what he calls "the howling cry of famished organs", which were "by passion undermined".[86] Such a depiction of the hunger of an invalid voices a timeless and complex lived experience that was informed in this period by a nexus of discourses in literature, gastric biology, neurology and the burgeoning genre of dietetics. As the "Dyspeptic Saints" article argued, to be critically unmindful of the "philosophy of the stomach"[87] was to be insensitive to the sometimes indistinguishable boundary between physiological and psychological distress—or, between being hungry and being angry.

[85] Oudenhove et al., "Psychosocial Factors, Psychiatric Illness and Functional Gastrointestinal Disorders," 203–204.

[86] "Hunger", *The London Journal*, 424.

[87] "Dyspeptic Saints," 26.

Bibliography

A Hypochondriac, "What to Eat, Drink, and Avoid". *The City Jackdaw* 1 (15 September 1876): 379.

"A Physicians Notebook". *The Critic* 19, no. 483 (8 October 1859): 342–344.

Abernethy, John. *Surgical Observations, Part the Second; Containing an Account of Disorders of the Health in General and of the Digestive Organs in Particular.* London: Longman, Hurst, Rees, and Orme, 1806.

[Advertisement]. *The Athenaeum* 1118 (31 March 1849): 343.

Alberti, Fay Bound. *The Mortal Coil: The Human Body in History and Culture.* Oxford: Oxford University Press, 2016.

Alexander, Philip. *On the Treatment of a More Protracted Case of Indigestion.* London: T. and G. Underwood, 1827.

"ART. II.-The Infirmities of Genius Illustrated by Referring the Anomalies of the Literary Character to the Habits and Constitutional Peculiarities of Men of Genius". *The Quarterly Review* 50, no. 99 (October 1833): 34–56.

Beard, George Miller. "The Extent of Insanity". *Dublin University Magazine,* 79, no. 474 (June 1872): 668–699.

"Beaumont's Experiments on Digestion". *Chambers's Edinburgh Journal* 342 (18 August 1838): 234–235.

Beaumont, William. *Experiments and Observations on the Gastric Juice and the Physiology of Digestion*, edited by Andrew Combe. Edinburgh: MacLachlan and Stewart, 1838.

Bell, Robert. "On the Treatment of a Stomach". *The Butterfly* 2 (September 1899): 135–139.

Bohshedt, John. "Food Riots and the Politics of Provisions in World History". *IDS Working Paper,* 444 (2014), Brighton: IDS, 13 February. http://onlinelibrary.wiley.com/store/10.1111/j.2040-0209.2014.00444.x/asset/idwp444.pdf?v=1&t=jbxng3ao&s=b08426155e18a3b75c025da6d07ba571b3d1e81a.

Bown, Nicola. "What Is the Stuff That Dreams Are Made Of?". In *The Victorian Supernatural,* edited by Nicola Bown, Carolyn Burdett, and Pamela Thurschwell, 151–172. Cambridge: Cambridge University Press, 2004.

Boyce, Charlotte. "Representing the 'Hungry Forties' in Image and Verse: The Politics of Hunger in Early-Victorian Illustrated Periodicals". *Victorian Literature and Culture* 40 (2012): 421–449.

Broussais, F. J. V. *Principles of Physiological Medicine in the Form of Propositions, Embracing Physiology, Pathology, and Therapeutics, With Commentaries on Those Relating to Pathology.* Philadelphia: Carey and Lea, 1832.

Cheyne, George. *The English Malady; or a Treatise of Nervous Diseases of all Kinds as Spleen, Vapours, Lowness of Spirits, Hypochondriacal and Hysterical Distempers & etc.* London: G. Strahan and J. Leake, 1733.

[Classified Ad 29]. *Manchester Guardian* (2 October 1897): 4.

Colburn, Glen. "Introduction." In *The English Malady: Enabling and Disabling Fictions*, edited by Glen Colburn. Newcastle: Cambridge Scholars Publishing, 2008.

Coombe, George. *A System of Phrenology*, 5th ed., 2 vols. Edinburgh: Neill and Co, 1853.

Dickens, Charles. *The Chimes: A Goblin Story*. 1844; CreateSpace, 2016.

———, ed. "Stomach and Heart". *All the Year Round* 17, no. 419 (4 May 1867): 438–440.

———. *Great Expectations*. Boston: Estes and Lauriat, 1881.

"Dyspepsia". *Bow Bells: A Magazine of General Literature and Art for Family Reading* 14, no. 3 (1871): 54.

"Dyspeptic Saints". *St James's Magazine*, 2 (October 1868): 24–32.

Finlay, Mark. "Quackery and Cookery: Justus von Liebig's Extract of Meat and the Theory of Nutrition in the Victorian Age". *Bulletin of the History of Medicine* 66, no. 3 (1992): 404–418.

First, Michael B., and Allan Tasman. *DSM-IV-TR Mental Disorders: Diagnosis, Eitology, and Treatment*. Chichester: Wiley, 2004.

Ford, Jennifer. *Coleridge on Dreaming: Romanticism, Dreams, and the Medical Imagination*. Cambridge: Cambridge University Press, 1998.

French, Michael, and Jim Phillips, eds. *Cheated Not Poisoned?: Food Regulation in the United Kingdom, 1875–1938*. Manchester: Manchester University Press, 2000.

Gregory, James. *Of Victorians and Vegetarians: The Vegetarian Movement in Nineteenth-Century Britain*. London: Tauris Academic Studies, 2007.

Gurney, Peter J. ""Rejoicing in Potatoes": The Politics of Consumption in England During the Hungry Forties" *Past and Present* 203, no. 1 (1 May 2009): 99–136.

Hall, S. C., ed. "Molly's Dream". *Sharpe's London Magazine of Entertainment and Instruction for General Reading* 33 (July 1868): 91.

"Hangry". Dictionary.com. *Dictionary.com Unabridged*. Random House, Inc. http://www.dictionary.com/browse/hangry. Accessed 26 September 2017.

Harkness, Margaret. *In Darkest London*. 1889; Black Apollo Press, 2003.

Hassall, Arthur Hill. *Food and Its Adulterations, Comprising the Reports of the Analytical Sanitary Commission of "The Lancet" for the Years 1851–1854 Inclusive*. London: Longman, Brown, Green, and Longmans, 1855.

———. *Adulterations Detected, or Plain Instructions for the Discovery of Frauds in Medicine and Food*. London: Longman, Brown, Green, and Longmans, 1857.

Heath, Stephen. "Hypochondria: Medical Condition, Creative Malady". *Brain* 134, no. 4 (1 April 2011): 1244–1249.

Houston, Gail Turley. *Consuming Fictions: Gender, Class, and Hunger in Dickens's Novels.* Carbondale: Southern Illinois University Press, 1994.

Huff, Joyce L. "The Narrating Stomach: Appetite, Authority and Agency in Sydney Whiting's 1853 *Memoirs of a Stomach*". *Body Politics* 3, no. 5 (2015): 75–93.

"Hunger". *The London Journal, a Weekly Record of Literature, Science, and Art* 3, no. 80 (5 September 1846).

Hutson, David. J. "Plump or Corpulent? Lean or Gaunt? Historical Categories of Bodily Health in Nineteenth-Century Thought". *Social Science History* 41, no. 2 (2017): 283–303.

Ishizuka, Hisao. "Carlyle's Nervous Dyspepsia: Nervousness, Indigestion, and the Experience of Modernity in Nineteenth-Century Britain". In *Neurology and Modernity. A Culture History of Nervous Systems, 1800–1930*, 81–95, edited Laura Salisbury and Andrew Shail. Basingstoke: Palgrave, 2010.

James, John. "ART. I. An Essay on Indigestion--: Pathological Inquiries. Indigestion. Treatment of Indigestion". *The Philadelphia Journal of the Medical and Physical Sciences* 9, no. 17 (1824): 1–30.

Johnson, Edward. *Results of Hydropathy: Or, Constipation Not a Disease of the Bowels: Indigestion Not a Disease of the Stomach; with an Exposition of the True Nature and Cause of These Ailments, Explaining the Reason Why They Are So Certainly Cured at Stanstead Bury House, With Observations of the Treatment Generally.* New York and London: Wiley, 1849.

"Kutnow's Powder" [Advertisement]. *Review of Reviews* (September 1898): 7.

Le Fanu, J. Sheridan. "Green Tea" (1869). In *A Glass Darkly*, vol 1, 58. London: R. Bentley and Son, 1872.

Madden, Richard R. *The Infirmities of Genius Illustrated by Referring to the Anomalies of the Literary Character to the Habits and Constitutional Peculiarities of Men of Genius.* Philadelphia: Carey, Lea, and Blanchard, 1833.

Mitchell, John. *A Treatise on the Falsification of Food.* London: Hippolyte Bailliere, 1848.

Mother Seigel's Syrup [Advertisement]. *The Legate Illustrated Magazine*, 7 (September 1894): 2.

"On the Treatment of the More Protracted Cases of Indigestion". The Eclectic Review 27(May 1827): 405–427.

"One Volume Fiction". *The Speaker: The Liberal Review* 3 (4 April 1891): 413.

Pereira, Jonathan. *A Treatise on Food and Diet: With Observations on the Dietetical Regimen Suited for Disordered States of the Digestive Organs and an Account of the Dietaries of Some of the Principal Metropolitan and Other Establishments for Paupers, Lunatics, Criminals, Children, the Sick, &c.* New York: Fowler and Wells, 1843.

Porter, Roy. "Nervousness, Eighteenth and Nineteenth Century Style: From Luxury to Labour". In *Cultures of Neurasthenia from Beard to the First World*

War, 31–50, edited by Marijke Gijwijt-Hofstra and Roy Porter. Amsterdam: Rodopi, 2001.

Porter, Roy, and G. S. Rousseau. *Gout: The Patrician Malady*. New Haven: Yale University Press, 2000.

Salisbury, Laura, and Andrew Shail. "Introduction". In *Neurology and Modernity. A Culture History of Nervous Systems, 1800–1930*, edited by Laura Salisbury and Andrew Shail. Basingstoke: Palgrave, 2010.

Sampson, Henry, ed. "Indigestion! A Christmas Carol". *Fun* 28 (18 December 1878): 252.

Schoffer, Kerrie L., and John D. O'Sullivan. "Charles Dickens: The Man, Medicine, and Movement Disorders". *Journal of Clinical Neuroscience* 13 (2006): 898–901.

Scott, John, and John Taylor, eds. "Every Man's Master". *The London Magazine* 3, no. 13 (1829): 246–358.

Shuttleworth, Sally. *The Mind of a Child: Child Development in Literature, Science, and Medicine 1840–1900*. Oxford: Oxford University Press, 2014.

"The *BMJ* on the Death of Charles Dickens". *British Medical Journal* 344 (2012): 630.

"The Chimes: A Goblin Story". *Hood's Magazine and Comic Miscellany* 3, no. 1 (January 1845): 74–81.

"The Demon of Dyspepsia". *Judy: Or the London Serio-Comic Journal* 33 (26 December 1883): 307.

"The Literature of Spiritualism". *The Examiner* 3696 (30 November 1878): 1518–1519.

Thomson, Anthony Todd. "Extracts from the Note-Book of a Physician". *New Monthly Magazine* 57, no. 226 (October 1839): 193–201.

———. "Sketches from the Note-Book of a Physician, No. III". *New Monthly Magazine* 58, no. 230 (February 1840): 260–271.

Timbs, John. *Hints for the Table; or the Economy of Good Living*. London: Simpkin, Marshall, and Co, 1838.

van Oudenhove, Lukas, Joris Vandenberghe, Koen Demyttenaere, and Jan Tack. "Psychosocial Factors, Psychiatric Illness and Functional Gastrointestinal Disorders: A Historical Perspective". *Digestion* 82 (2010): 201–210.

von Liebig, Justus. *Animal Chemistry; Or Organic Chemistry in Its Application to Physiology and Pathology*, ed. William Gregory. Cambridge: John Owen, 1842.

Whiting, Sydney. *Memoirs of a Stomach; Written by Himself, That All Who Eat May Read*. London: John Murray, 1853.

CHAPTER 7

Visceralism and the Superior Mind in French Medicine and Literature, 1750–1850

Anne Vila

"I dare you to explain anything without the body": so declared Denis Diderot in his *Éléments de physiologie* (1778).[1] The corporeal side of human life was indeed vital to pre-Revolutionary French culture: somatic expression was central to theories of painting, acting, music, and language; the quest for physical contentment drove the growing market for consumer goods; and philosophers of the school known as *sensualisme* or sensationism gave a fundamental role to the impressions the mind received through the senses. Sensation, as Voltaire put it, "envelopes all of our faculties"[2]—reflecting a perspective that drove various theorists to devise metaphysical "anatomies" to analyze the particular

[1] Diderot, *Éléments de physiologie*, 334. All translations from the French are my own, unless otherwise noted.

[2] Voltaire, "Sensation," 530.

A. Vila (✉)
University of Wisconsin–Madison, Madison, WI, USA

M. Mathias and A. M. Moore (eds.), *Gut Feeling and Digestive Health in Nineteenth-Century Literature, History and Culture*, Palgrave Studies in Literature, Science and Medicine,
https://doi.org/10.1007/978-3-030-01857-3_7

mental capacities a person got from each of the senses he or she possessed.[3] Some authors also speculated about how thinking processes were affected by more internalized sensations like gut feelings.

Take, for example, Voltaire's article "Passions" of the *Questions sur l'Encyclopédie* (1774), where he asked: "Tell me what secret connection nature has created between an idea and a bowel movement?"[4] That question (directed, in its context, at a hypothetical theologian who insisted on the mind's immateriality) shows Voltaire's conviction that there was a crucial, albeit mysterious link between mental operations and visceral matters. More broadly, it illustrates what Daniel Cottom has called the French Enlightenment's "visceral turn," which reoriented epistemological inquiry "away from conjectural innate ideas and toward the palpable, dissectable, scrutable organs of the body."[5] This emphasis on the functions and effects of the lower organs coexisted with the rise of "nerve" theory for which this period is well known: as Rebecca Spang explains, "the processes of digestion (or their dread counterpart, indigestion) mediated much of the external world's impact on those nerves. Nerves [...] did not so much replace the stomach as the century's 'magical human organ' as they extended its domain."[6]

Multiple aspects of eighteenth-century French culture fostered interest in the gut's effects on the moral/mental realm. One was the aristocratic fad known as the vapors, a digestively related nervous ailment that inspired the genre of "flatulent" literature popular in some worldly circles.[7] Another was the invention of the restaurant, a development connected to the expansion of gastronomy and food science—and, more

[3] The term "metaphysical anatomy" comes from Diderot, *Lettre sur les sourds et muets*, in *Œuvres complètes*, vol. 4, 140. Other sensationist anatomies include Étienne Bonnot de Condillac's statue-man fable in *Traité des sensations* (1754); the chapter "De l'homme" in Georges Louis Leclerc, comte de Buffon's *Histoire naturelle* (1749–89); and Charles Bonnet, *Essai analytique sur les facultés de l'âme* (1760).

[4] Voltaire, "Passions," 377.

[5] Daniel Cottom, *Cannibals and Philosophers*, xii, 7.

[6] Rebecca L. Spang, *The Invention of the Restaurant*, 38.

[7] Marianne Charrier-Vozel illustrates the popularity of flatulent literature by citing a letter in which Mme d'Epinay described the therapeutic farting she used to relieve her colic: "Sociabilités de la maladie," 237.

broadly, to hygiene, a part of medicine that underwent a marked rise in France after 1700.[8]

A key task hygienists assigned themselves was to update the ancient doctrine of the non-naturals, which, classically defined, were air, food and drink, exercise and sleep, bodily retentions and evacuations, and the passions of the soul. Striking a healthful balance among the non-naturals required a regimen, and because regimen was largely a matter of diet before 1780, physicians exhorted laypersons to pay close attention to the influence that digestive operations exerted over mood, temperament, and intellectual endeavor.[9] They were particularly concerned about the patient group identified most closely with mental work: *gens de lettres*, a blanket term used at the time to describe everyone from poets to mathematicians.

This essay offers an overview of the role played by visceralism in the evolving image of the intellectual in French-speaking Europe from approximately 1750 to 1850. I will start by considering the rise of *les maladies des gens de lettres*, a disease syndrome whose central tenet was that intellectuals had a unique physiology that included weak, sensitive stomachs. I will then examine the role of visceralism in early nineteenth-century medical efforts to redefine the disorders known as hypochondria, melancholy, and hysteria; here, I will focus specifically on the figure of the male intellectual, whose identity became more firmly grounded in dyspepsia along with depressive psychic states. To conclude, I will look briefly at visceralist images of the intellectual in and around the work of the novelist and social theorist Honoré de Balzac, who echoed (somewhat facetiously) medical concerns about how the special physiology of cerebralists was affected by the ingestion of modern drinks like tea and coffee.

Visceralism and the *Maladies des Gens de Lettres* Syndrome

Visceralism is not a topic typically featured in historical explanations of the persona of the intellectual in eighteenth- and early nineteenth-century France. However, this period's views on thinking often diverged

[8]On restaurants and food science, see Spang, *Invention of the Restaurant*, and E. C. Spary, *Eating the Enlightenment*. On the rise of hygiene, see Laurence Brockliss and Colin Jones, *The Medical World of Early Modern France*, 441–473.

[9]As Spary emphasizes, literate Europeans "construed their eating habits in terms of regimen" (*Eating the Enlightenment*, 254).

from the transcendent, disembodied rationalism that is often associated with the Enlightenment movement. Despite the persistence of Cartesian substance dualism and Christian apologetic approaches to the soul, the mind was not neatly separated from the body by many Enlightenment-era mind theorists. To the contrary, these theorists saw moral, mental, and physical operations as deeply interconnected.[10] Moreover, they saw mental work as an activity that engaged the passions, which they typically located in the lower, visceral realms of the human being. The intellectual was, in short, very much a flesh-and-blood creature, a being whose internal parts and operations elicited interest in many fields.

The so-called Age of Reason, when intellectuals enjoyed unprecedented powers and influence throughout Europe, was also the era that gave birth to the nosological category of illnesses proper to the intelligentsia. Two themes pervaded the works physicians wrote to define the distinctive traits of this new class of pathologies. The first was the idea that scholars were more susceptible to disease than most people because of the physical strains caused by excessive use of the mind. The second was a mixture of fascination and perplexity in the face of the passion that drove certain people to sacrifice their health to the relentless pursuit of knowledge. Both lay the groundwork for the emergence of later jointly cultural and moral syndromes like monomania and obsession, and both contributed to the creation of the various "sick heroes" who became prominent in European culture—including the figure of the dyspeptic artist that, as Bertrand Marquer has argued, was both a medical norm and an aesthetic anti-model in the nineteenth century.[11]

Historians generally trace the rise of medical concern over intellectual pathologies to Bernardino Ramazzini's *De Morbis Artificium diatriba* (1700) translated into English in 1705 as *Treatise of the Diseases of Tradesmen*, which concluded with a long chapter on "The Diseases of Learned Men." Ramazzini presented such diseases as the product of

[10] See Fernando Vidal, *Les sciences de l'âme*; and Gary Hatfield, "Remaking the Science of Mind."

[11] See Bertrand Marquer, "Portrait de l'artiste en dyspeptique." On monomania and obsession, see, among other studies, Jan Goldstein, *Console and Classify*, 152–196. Studies of the nineteenth-century writer or artist as a "sick" hero include Alan Pasco, *Sick Heroes*; and Myriam Roman, "Avatars romanesques du penseur chez Mme de Staël, Balzac et Hugo."

professional circumstances rather than innate temperament.[12] He singled out the sedentariness and bad posture of studious people: they were "as Slothful and Idle in their Body, as they are active in their Mind and Brain" (247), and spent too much time hunched over books. This was to blame for the weak stomachs they typically suffered (an ancient notion, first popularized by Aulus Cornelius Celsus), along with their susceptibility to compressed pancreatic juices and "Nephrtick and Athritick Disorders" (250). By classifying learned people in terms of the bodily effects of their labor, Ramazzini inspired physicians to extend their attention beyond that group's diet (an approach already well established in the European medical and moralist tradition) into other areas of the scholarly life and work routine.

The first wave of physicians to heed Ramazzini's call for closer attention to the working conditions of the studious was the English "hyp" doctors of the 1720s and 1730s, like Bernard Mandeville, Nicholas Robinson, and George Cheyne.[13] In the French context, early adapters were the Jansenist physician Philippe Hecquet, who devoted a chapter to the diseases of the learned in *La Médecine, la chirurgie et la pharmacie des pauvres* (1740),[14] and the fashionable Parisian doctor Anne-Charles Lorry, who included a special fifteen-page section on the dietary needs of men of letters in the second, expanded edition of his *Essai sur l'usage des alimens* (1757; 1st ed., 1754). Lorry placed *gens de lettres* in the second of the two categories of "necessary" people who deserved extra attention from physicians (in the first were soldiers and other muscular sorts, who served as "the arms of the State"[15]). His general premise was that "one must start early in thinking about one's body, observing exactly the proportion that must exist between the sum of one's excrements and that of one's food" (II, 210–11). However, those who spent the most time cogitating were precisely the people least apt to spend much time thinking about their bodies. Moreover, study and meditation constituted "a sort of internal life that unfolds at the expense of the external life, harming

[12] Ramazzini, *Treatise of the Diseases of Tradesmen*, 246. Other citations to this work appear parenthetically in the text.

[13] See George S. Rousseau's introduction to John Hill, *Hypochondriasis: A Practical Treatise, 1766*; German E. Berrios, "Hypochondriasis: A History of the Concept," and "In the Name of Hygieia and Hippocrates: A Quest for the Preservation of Health and Virtue."

[14] See Dinah Ribard, "Pathologies intellectuelles et littérarisation de la médecine."

[15] Lorry, *Essai sur l'usage des alimens*, vol. 2, 216.

the function of the machine" (II, 235). Although Lorry urged his scholarly readers to remember that "the first attention of a *Philosophe* must be not to work nor meditate while his stomach is digesting" (II, 237–38), he was not optimistic that they would follow his advice.

From mid-century on, doctors on the continent published a steady stream of works in the vernacular that addressed the physical and moral effects of study. A few were reassuring, like *La Médecine de l'esprit* (1753) by the Parisian Doctor-Regent Antoine Le Camus, who offered precise dietary and hygienic regimens on how to reach and maintain one's full intellectual potential. Other works were alarmist, like those produced by two Swiss doctors, Johann-Georg Zimmermann and Samuel-Auguste Tissot. Zimmermann included a long chapter on the effects of excessive mental exertion in *Von der Erfahrung in der Arzneikunst* (1763, translated into French as *Traité de l'expérience en médecine* in 1774), where he spoke of the love of learning as a noble but dangerous passion that made scholars susceptible to digestive disorders, incapacitating headaches, weakened nerves, loss of vision and hearing, and hypochondria (a disease defined as physical as well as moral, and related to melancholy[16]). Tissot devoted an entire treatise to the question: *De la santé des gens de lettres*, first published in 1768 and significantly expanded in its third edition (1775). Filled with frightening tales of people who had destroyed their health through excessive mental application, this book was read by some of the most illustrious members of the European intellectual elite.

While acknowledging those who had written before him on the topic, Tissot claimed to be the first physician to offer a systematic account of the health circumstances "that differentiate the state of Scholars from that of other orders of society."[17] He offered arresting portraits of scholars laid low by overstudy, intermingled with frequent appeals to the culture of his readers, interweaving his own ideas about the causes of illness in scholars with those of his medical contemporaries and notions that had been circulating since antiquity. Tissot's list of

[16] "Hypochondriacal (passion or affection): This is the term ordinarily used by physicians to designate a species of malady related to melancholy, since atrabile is also its morbific humor, which infects the entire mass of fluids […] but settles particularly on the organs or viscera of the lower abdomen" ("Hypochondriaque," in *Encyclopédie, ou dictionnaire raisonné des sciences, des arts et des métiers*, vol. 8, 408–409).

[17] Tissot, *De la santé des gens de lettres*, xii. Abbreviated here as *SGL*.

possible study-induced disorders included everything from poor digestion to more dire conditions like the sensory impairment observed in the renowned Greek scholar Madame Dacier, who, "while reciting Hector's farewell to Andromache, was so deeply moved that she lost the use of her senses" (*SGL*, 19).

The thematic structure of *De la santé des gens de lettres* follows the standard categories of medical hygiene: the book covers all the ways in which scholars used and misused the six non-naturals.[18] The topics of food and drink take up a good fifty pages of the book (articles 57–73, out of a total of 92). Among other things, Tissot took scholars to task for their hasty eating and inadequate chewing (*SGL*, 173), and for their tendency to defer urination and defecation, a bad habit he attributed to "the ardor for work taken to a ridiculous and blameful excess that does not allow a person to take the time to eat or drink" (97). Finally, he warned scholars against their tendency to overeat. For Tissot and like-minded doctors, appetite was "the somatic locus where powers of mind might be overwhelmed by bodily desires."[19]

Much of the dietary advice Tissot offered to intellectuals was based on the ancient maxim that "the man who thinks the most is the one who digests the most poorly" (*SGL*, 24), which he cited to make a backhand compliment to his learned readers. What is significant about Tissot's perspective on the scholarly stomach was the manner in which he modernized it. First, he drew on the idea–promoted by the influential vitalist theorists of the Montpellier medical faculty, where he had done his training–that the abdomen and brain were rival centers of physiological function: one center reacted antagonistically when the other center monopolized the body's vital energies.[20] Second, he viewed contemporary *gens de lettres* as trend-followers, caught up in both the fad of learning and in gastronomic fads like tea-drinking.

[18]See Antoinette Emch-Dériaz, "The Non-Naturals Made Easy." See also her discussion of *La Santé* in *Tissot, Physician of the Enlightenment*, 72–77.

[19]Spary, *Eating the Enlightenment*, 250.

[20]The theory of vital regions owed much of its success in the eighteenth century to the prominence of Montpellier vitalism, one of whose trademark ideas was what Elizabeth Williams has called a "triadic conception" of the body (Williams, *The Physical and the Moral*, 38–39, 43–44, 51, 59). See also Roselyne Rey, *Naissance et développement du vitalisme en France*, 164–169.

Despite the many pages Tissot devoted to diet and recreation, the most important category of the non-naturals in *De la santé des gens de lettres* is the sixth one, the passions of the soul. The passions held dangers for scholars not simply because they were so obstinately attached to their studies, but also because their mental overwork made them susceptible to intense, irrational fears. That is the theme of the most dramatic tales Tissot recounted, like the vignettes of Tasso, Spinello, Pascal, and other famous thinkers, writers, and artists who went totally or partially insane (*SGL*, 44–51). The chain of causation in their pathologies started in the mind, proving the general principle that "the mind long occupied has impressed too strong an action upon the brain and is no longer able to suppress it" (43). In physical terms, prolonged mental application was like a "ligature" applied to all of the nerves at once; when the mind upset the organism to such an extent, the results were catastrophic.

Tissot's moralizing rhetoric is most pronounced in the article that marks the midpoint of *La santé des gens de lettres*: article 52, where he shifted from the grim recounting of ailments induced by overstudy into an explanation of the practical measures by which scholars could restore and maintain their health. Noting that *gens de lettres* were typically difficult patients to treat, he compared them to "lovers who fly off the handle when one dares to say that the object of their passion has defects; moreover, they almost all have the sort of fixity in their ideas that is created by study" (*SGL*, 132). In the next article, he likened them to selfish gluttons: "most of them don't even have the public in mind, and devour study only as the gourmand devours meat: to assuage his passion. [...] Rush them, drag them out of their *cabinet*, force them to engage in rest and entertainments that will ward off ills and restore their strength" (134–135).

Tissot's fundamental advice was this: to insure their health and longevity, scholars needed to exercise moderation in their mental labors. That meant avoiding "literary intemperance" (*SGL*, 28), a term he used to refer to any sort of overly assiduous reading or studious reflection—and which he expressly linked to digestive as well as sensory disorders. Intellectuals *could*, he stressed, cultivate their chosen specialty if they interrupted their studies often enough to restore balance in their lives and in their bodies (135). At bottom, it all came down to sobriety, broadly construed: *gens de lettres* had to learn to control their extraordinary appetites for learning.

Despite its frightening scenarios (and occasional echoes of Jean-Jacques Rousseau's famous diatribes about the ills of book-learning[21]), Tissot's treatise was enthusiastically received by many contemporary intellectuals. Charles Bonnet told Tissot that he had felt "goose bumps from head to feet" while reading it, and complained that he had not yet succeeded in "evangelizing" their mutual friend Albrecht von Haller (a notorious workhorse) on the subject of the hygienic principles Tissot had established for the scholarly regimen.[22] It also got an avid reception from some of Tissot's patients, who were eager to see their own symptoms reflected in the pages of his book.[23] Voltaire, for his part, was dismissive of Tissot's zeal for medical popularization[24]; but he nonetheless adhered to an activist system of hygiene and dietetics—including regular self-purging to keep his mind and bowels clear—that was reminiscent of the ideas compiled in *De la santé des gens de lettres.*[25] Less than ten years after its first French edition, this book had already been translated into English, German, Italian, Spanish, and Polish, and it would continue to be reprinted regularly as a health manual until 1859.

Visceralism and the Gendering of Nervous Disorders

Visceralism also marked the fields of mental medicine and anthropological medicine in Revolutionary and post-Revolutionary France—both of which took over part of the evolving discussion of "learned" ailments. From the pioneering psychiatrist Philippe Pinel to the medical Idéologue Pierre-Jean-George Cabanis, doctors were intent on teasing apart the

[21] Tissot cited Rousseau's "Préface de Narcisse" to support his contention that library work debilitated the body (*SGL*, 31) but ultimately took a cautious distance from Rousseau's polemical condemnation of the arts and sciences. See on that question, Ronan Chalmin and Anne Vila, "Malade de son génie: raconter les pathologies des gens de lettres, de Tissot à Balzac."

[22] Cited in Miriam Nicoli, *Les savants et les livres*, 145.

[23] See Séverine Pilloud, *Les mots du corps*, 279–281.

[24] Voltaire called Tissot a "fabricator of little medical books" in a 1772 letter to the Marquise du Deffand (*Œuvres complètes de Voltaire. Correspondance générale*, vol. 56, 71).

[25] For a fuller discussion of Voltaire's personal regimen—and visceralism in his satirical writings—see my essay, "The Philosophe's Stomach."

previously intertwined diseases known as hypochondria, melancholy, and hysteria. One means by which they did so was to locate these conditions differently in the lower body.

Enlightenment-era physicians had spoken more or less interchangeably of "nervous," hypochondriac," and "hysteric" disorders, a trio of adjectives featured together in a 1765 treatise by the English nerve doctor Robert Whytt, and they generally agreed with the rapprochement of hysteria and hypochondriasis that had been proposed by the seventeenth-century theorists Thomas Willis and Thomas Syndenham—an alignment that made the two disorders "sibling syndromes with similar symptomatologies that traced to the nervous system."[26] That mixed concept (which also included melancholy) was abandoned after 1790, around the same time that a new theory of sexual incommensurability set in across Europe.

A good deal of historical work has been done on the ways in which hysteria was separated out from its former "sibling syndromes" and redefined as a properly *female* disease at the end of the eighteenth century. One of the best-known contributions to this endeavor was *Recherches historiques et médicales sur l'hypocondrie, isolée, par l'observation et l'analyse, de l'hystérie et de la mélancolie* (1802) by Pinel's associate Dr. Jean-Baptiste Louyer-Villermay, who refeminized hysteria by reviving the age-old uterine theory of its origins.[27] Less attention has been paid to the efforts made by Villermay and other early French alienists to redefine hypochondria as a male disorder—and to associate it particularly with male intellectuals.

These doctors were not staunchly mentalist in regard to the psychic disorders that they perceived as stemming from intellectual exertion: many insisted that such disorders had their seat in an organic lesion, usually in the stomach.[28] That approach was consistent with the perspective of predecessors like Tissot, who, in *De la santé des gens de lettres*, divided hypochondria into two species: "the kind that is simply nervous

[26] Mark S. Micale, *Hysterical Men*, 18–19. See also his overview of the late eighteenth-century shift in medical thinking about nervous disorders (49–116).

[27] On Villermay and the gendering of hysteria, see Étienne Trillat, *Histoire de l'hystérie*, 101–107; Nicole Edelman, *Les Métamorphoses de l'hystérie*, 16–22; and Sabine Arnaud, *On Hysteria*, 27–28 and passim.

[28] See Elizabeth A. Williams, "Stomach and Psyche."

and ... is the effect of mental exertion," versus the kind "that depends on the engorgement of the viscera of the lower abdomen and digestive disturbance, which always results from inaction. It is easy to understand how these two sorts of illness are found in combination among Men of Letters. It is rare that they not be affected by them to some degree, and so difficult to cure them entirely" (*SGL*, 76). Tissot's suggested therapies to counter hypochondria also interwove the moral and the physical. This flexible perspective persisted in early nineteenth-century accounts of the illness, which often cited symptoms of visceral discomfort along with psychological conditions like taciturn moroseness. Villermay, for example, warned that hypochondria could be induced by undertaking mental work too soon after eating, or by eating and thinking at the same time.[29]

Hypochondria, as these doctors portrayed it, involved worrisome emotional states as well as abdominal symptoms like constipation, slow digestion, and flatulence. Pinel tied hypochondriacal disorders to sadness and capriciousness of character—sometimes aggravated by changes in the atmosphere, as well as by contemplative life—whereas Villermay proposed a longer list of symptoms that included panic attacks and exaggerated anxieties like fretting about one's health.[30] Melancholy was held to involve more violent, somber, obsessive passions, including a stubborn, sometimes delirious fixation on a single idea. It was also more closely linked than hypochondria to a particular temperament, one that these doctors claimed was most common among dark-haired men with slow gaits and perpetually sad demeanors.[31] Yet hypochondria could, they warned, degenerate into melancholy, especially when the viscera remained in a chronic state of irritation due to overstudy, habitual overeating, or other causes.[32]

Well into the 1820s, French doctors maintained that the seat of hypochondria resided in the abdomen, not the brain. In the most radical expression of this view, the controversial medical theorist

[29] Villermay, *Recherches historiques et médicales sur l'hypocondrie*, 373.

[30] Pinel, *Nosographie philosophique*, vol. 2, 8–10; Villermay, *Traité des maladies nerveuses ou vapeurs*, vol. 1, 223.

[31] See, for instance, Villermay, *Recherches historiques et médicales sur l'hypocondrie*, 60.

[32] See, for example, C. A. T. Charpentier, *Essai sur la mélancolie*, 62–63. As Susan Kassouf points out, late eighteenth-century European doctors also attributed those symptoms and habits to the Jewish male body ("The Shared Pain of the Golden Vein").

François-Joseph-Victor Broussais declared that it was all a matter of gastritis.[33] Locating hypochondria and its sibling syndromes in the viscera served both practical and theoretical purposes. First, it provided physicians with a justification for prescribing body-based treatment options like exercise and manual labor, purgatives and other digestive remedies, and hydrotherapy. Those therapies persisted despite the introduction of Pinel's famous moral treatment, which aimed to cure mental illness either by distracting the patient or by shaking up his or her imagination in order to break the "vicious chain of ideas" that had taken hold of it.[34] Second, the visceral location of these disorders adhered to the triadic conception of human physiology that had been advanced by the Montpellier medical vitalists, who regarded the abdomen as one of the body's three vital centers. Finally, it provided a handy anatomical framework for drawing gender distinctions between male versus female sufferers. Following that logic, French doctors tended increasingly to reserve the terms "vaporous" or "hysterical" for women, on the theory that they were suffering from uterine spasms, and to call men "hypochondriacal." To a degree, therefore, hypochondria made it possible for turn-of-the-century *gens de lettres* to be nervous in a manly way—and somewhat more curable than the patient groups to which they were set in opposition.

Cabanis also presented a normative image of the scholarly constitution in the *Rapports du physique et du moral de l'homme* (1802), an influential contribution to early nineteenth-century medical and social theory. Cabanis's book was controversial in some French intellectual circles because of the comparison he made between the stomach and the brain in his remark "the brain digests impressions, as it were, and produces organically the secretion of thought."[35] Yet viscera interested Cabanis for reasons that extended beyond that notorious analogy: he saw visceral weakness not just as a consequence of overstudy but also as a prerequisite

[33]Broussais, *De l'irritation et de la folie*, 645. Jackie Pigeaud refutes the "epigastric metaphor" proposed by some historians of psychiatry by emphasizing that Broussais's epigastric theory of passions and mental states was literal and materialist (*Aux portes de la psychiatrie: Pinel, l'ancien et le moderne*, 282).

[34]Goldstein, *Console and Classify*, 93.

[35]Cabanis, *Rapports du physique et du moral de l'homme*, 138; Cabanis, *On the Relations Between the Physical and Moral Aspects of Man*, vol. 1, 117, translation modified (as are all translations of later citations from this work).

for intellectual talent. As he argued, "The greatest aptitude for work that requires either a strong, active imagination, or persistent and profound meditations, often depends on a generally ill state introduced into the system by the disturbance of the functions of certain abdominal organs."[36]

To appreciate the implications of that claim, we should consider the context in which it was made: memoir 5 of the *Rapports du physique et du moral*, where Cabanis addressed the influence of sex differences on the character of ideas. As he described it there, dyspepsia was a specifically *masculine* trait, one analogous to—but not conflatable with—the generally weak, excitable, capricious state created in the female system by the functions of the uterus and ovaries. Chronic visceral disturbance, in other words, had different consequences for each sex. It was a constitutional factor that enabled men of a certain temperament to excel in the intellectual realm, even as it exposed them to hypochondriacal nervous affections. For women, by contrast, this sort of disturbance was directed downward toward their reproductive organs, whose eminent sensibility was accompanied in the female body by a general softness that extended to the "cerebral pulp."[37] Women were simply not capable of deep thinking, in Cabanis's estimation: the healthy woman, although naturally intelligent, was "rightly put off by intellectual work that can't be carried out without long and profound meditations"; she preferred lighter subjects that required no more than a lively wit and imagination.[38] In response to reports that nervous ailments gave some women patients enhanced intellectual powers—as, for example, in bouts of vapors—he insisted that the effect was temporary: "nothing is less rare than to see them acquire during their vaporous attacks a penetration, a wit, an elevation of ideas and eloquence that they didn't have; those advantages, which are simply the result of the illness, disappear when health is restored."[39] He returned to this question in memoir 7, on the influence of illness on ideas, where he alluded to the claims made by magnetizers like the Lyon doctor Jacques-Henri-Désiré Petetin (author of the

[36]Cabanis, *Rapports*, 226; *Relations*, vol. 1, 224.

[37]Cabanis, *Rapports*, 229–230; *Relations*, vol. 1, 227. For a recent discussion of Cabanis's views on female constitution, see Florence Lotterie, *Le Genre des Lumières, Femme et philosophe au XVIIIe siècle*, 281–299.

[38]Cabanis, *Rapports*, 242; *Relations*, vol. 1, 241.

[39]Cabanis, *Rapports*, 248; *Relations*, vol. 1, 248.

Mémoire sur la découverte des phénomènes que présentent la catalepsie et le somnabulisme, 1787) that female cataleptics exhibited symptoms like heightened intelligence and the ability to see and even smell through their internal organs. Cabanis rebutted by declaring that "catalepsies, ecstasies, and all states of exaltation that are characterized by ideas and an eloquence that rise above the education and habits of the individual, are usually due to spasms in the organs of generation"—which is why those illnesses and the enhanced intelligence they sometimes produced were observed "mainly in women."[40]

By contrast, the enhancement effect was permanent in men who had an innately melancholic temperament or who had acquired a chronically dyspeptic state through force of habit. As he explained in memoir 7 of the *Rapports*, nervous affections in melancholic men "are conducive to attention and meditation; they accustom the senses and the organ of thought to exhaust, in a sense, the subjects of study to which they are attached. They expose the individual to all the errors of the imagination; but they can also enrich the genius with several precious qualities; they often lend talent a great deal of elevation, strength, and brilliance."[41] Cabanis thus gave an intriguing spin to the scholar's age-old "bad stomach" complaint: he interwove it with the sexually dimorphic model of human nature popularized by Rousseau and Dr. Pierre Roussel (author of *Système physique et moral de la femme*, 1775), while also linking it to the theory of limited perfectibility to which he, like the contemporary biomedical theorist Xavier Bichat, subscribed. Seen from this perspective, visceral sickliness in men was not the result of but the enabling mechanism of sustained, intense mental labor.

Dangerous Modern Stimulants, from Tea and Coffee Drinking to Excessive Thinking

Deterministic perspectives on the scholarly constitution were also advanced by two of Tissot's early nineteenth-century imitators: Étienne Brunaud, author *of De l'hygiène des gens de lettres* (1819), and Joseph-Henri Réveillé-Parise, whose *Physiologie et Hygiène des hommes livrés aux travaux de l'esprit* was first published in 1834 and reprinted regularly

[40] Cabanis, *Rapports*, 313; *Relations*, vol. 1, 328–329.

[41] Cabanis, *Rapports*, 314–315; *Relations*, vol. 1, 330.

until 1881. These armchair physicians lavished praise on intellectuals for perfecting the human spirit through their mental labors. However, they also warned about the worrisome side effects of those labors, going so far as to argue that repeatedly exerting one's mind creates a peculiar bodily conformation. Using a curious sort of analogical thinking, Brunaud compared the brain to three lower internal organs: the stomach, the bladder, and the uterus. Just as those organs acquired an exceptional degree of strength and volume when used habitually—as in men inclined to overeat, people with the "vicious" habit of retaining their urine for a long time, and women who were frequently pregnant—so, too, the scholarly brain expanded in size and power "through the effect of long and habitually sustained exercises and efforts."[42] Réveillé-Parise, for his part, reconfigured the frail musculature, difficult stomach, and exquisite sensitivity of those devoted to mental work, casting them as positive qualities: such "poetically organized" beings had reached a higher phase in the evolution of the human race toward greater perfection—a state demonstrated by the size of their heads, enlarged through the disproportionate concentration of vital activity in the brain and nerves.[43] What concerned these doctors more than big head size were the potentially life-shortening aspects of the nervous ailments and digestive disorders produced by "the cerebral excitation produced by sustained intellectual activities" (Brunaud, *De l'hygiène des gens de lettres*, 85–86). Because *gens de lettres* tended to seek out such excitation, they had to be extremely careful in regard to the non-naturals, from air and diet to the secretion of seminal fluid. Like Tissot before them, Brunaud and Réveillé-Parise proposed to guide their scholarly readers toward better, more healthful habits, ranging from how to eat, to what sorts of women to frequent, to ways to manage the intense stimulations and demands of modern life.

One finds a similar preoccupation with the glories and dangers of intellectual work in the writings of Honoré de Balzac, most particularly his *Études philosophiques* cycle, a set of novels designed to show

[42] Brunaud, *De l'hygiène des gens de lettres*, 37. Brunaud's belief that great minds were housed in big skull echoed other strands of early nineteenth-century scientific discourse; see Michael Hagner "Skulls, Brains and Memorial Culture."

[43] Joseph-Henri Réveillé-Parise, *Physiologie et hygiène des hommes livrés aux travaux de l'esprit*, vol. 1, 104.

the potentially fatal effects of excessive thinking or feeling.[44] Yet Balzac sometimes took a more wry approach to the topic, as in his pseudo-scientific *Traité des excitants modernes* (1839). He began this work by passing a general judgment on the consuming habits of his day: "The absorption of five substances discovered over the past two hundred years, and introduced into the human economy, has in recent years taken on such excessive proportions that modern societies find themselves modified in inestimable ways."[45] Balzac's pronouncement summed up decades of worry over the consumption of modern stimulants—a category that, as he formulated it, consisted of *l'eau de vie*, sugar, tobacco, tea, and coffee. In this vision of modernity, the act of ingestion was rife with risks that went beyond its effects on the individual human body: an entire society could be reshaped, for good or for bad, by the substances that its inhabitants chose to put into their bodies.[46] Moreover, the modern stomach was singularly liable to be overstimulated, because society had evolved toward greater civilization—the conditions that Balzac held to be most conducive to excess.

Although Balzac's treatise included reflections on chocolate, inebriation, the recent "invasion" of cigar smoking in France, and the potentially ghastly effects of overconsumption of sugar, its most piquant comments concerned the hot beverages on his list. Tea and coffee, he declared, had done the most to alter both the modern social economy and individual physiologies. As the English experience had proven, tea made women pale, sickly, boring, garrulous, and preachy (*TEM*, 320), whereas coffee was an "internal roaster" that had greatly accelerated the creative process for contemporary artists—like the composer Rossini and Balzac himself—but at the cost of stimulating them so much that they could be dangerous to be around (*TEM*, 315–316). Coffee was both particularly attractive and particularly pernicious to those whom he

[44] In the "Avant-Propos" to *La Comédie humaine*, Balzac explained that the general goal of the *Études philosophiques* was to show the "ravages of thought … sentiment by sentiment" (*La Comédie humaine*, vol. 1, 19). He also drew on Cabanis's "secretion" model of thinking in *Louis Lambert* (1832).

[45] Balzac, *Traité des excitants modernes*, 306; abbreviated here as *TEM*.

[46] Bertrand Marquer uses Balzac's novel *Le Cousin Pons* to support a suggestive parallel between the physiological concept of the "second stomach" and Balzac's image of the bourgeoisie as a median entity in the social body ("De l'épigastre au ventre").

addressed as "you illustrious Human Candles–you who consume your own brilliant selves with the heat and light of your minds" (316). It was thus at the top of Balzac's ranking of "most dangerous" modern stimulants, and it enjoyed that status because of its effects on a very particular set of modern consumers: Balzac and his fellow cerebralists.

Although Balzac did not cite Tissot in the *Traité des excitants modernes* (or elsewhere, for that matter), he was familiar with contemporary medico-hygienic doctrines and dropped references to Bichat, Cabanis, and Broussais in several of his novels.[47] It is therefore worth cycling back to Tissot to see why, back in the 1760s and 1770s, he had sounded the alarm about the rising popularity of tea and coffee among *gens de lettres*.

Tissot blamed the trend on an error that had crept into medicine after the discovery of circulation in the seventeenth century. Over-enthused by that discovery, some doctors had told their patients to drink huge quantities of warm stimulating beverages both day and night, on the theory that it would cure the thick blood from which they were all suffering. This practice had spread through Northern Europe with astonishing speed, triggering "a grievous, pronounced revolution in the history of health" and persisting particularly among studious people, whom Tissot held to be "naturally weak and feeble" (*SGL*, 191). Warm beverages, he warned, stretched and weakened the stomach's fibers, drowned the digestive juices, drove half-digested aliments to the intestines, and destroyed the fine mucus lining of the stomach and related viscera, making them oversensitive to the foods passing through them. Tea, having been "constantly imported from China and Japan [...] for these two centuries past," had multiplied the occurrence of "malignant and chronic disorders, especially nervous maladies"—propagating hypochondria, "sadness and despair" (*SGL*, 193–197).[48] Tissot was less censorious about coffee: "it raises the spirits, eases the stomach when loaded with phlegm, cures headaches, causes a cheerfulness of mind, and if we may believe some people, increases its penetration, for which reason the

[47] Moïse Le Yaouanc provides a brief but incisive overview of how Balzac's notion of the "nocivité de la pensée" draws on contemporary medical discourse, in *Nosographie de l'humanité balzacienne*, 47–49.

[48] Tissot's English translator, James Kirkpatrick, politely objected to his condemnation of tea drinking, calling it a product of Tissot's Swiss patriotism rather than his medical judgment; *An Essay on Diseases*, 149.

learned are so fond of it" (*SGL*, 201). *Gens de lettres* should, however, avoid consuming it every day: Tissot advised that they use coffee as a remedy (mixed with milk), not as part of their daily diet—and reminded them that great writers such as Corneille and Molière had produced their masterpieces without drinking coffee at all (202).

Balzac's semi-serious alarmism over coffee drinking by "Illustrious Human candles" clearly echoes elements of the discourse that French-language doctors had been producing for decades about the unique psycho-physiology of cerebralists.[49] A certain idea of perfectibility underpinned that medical advice: those who proffered it during Balzac's era believed that the future of civilization depended on the health and longevity of its intellectual luminaries. At the same time, they portrayed modern-day intellectuals as living in a state of mental intensity that was as devastating as it was sublime: although it was the motor of civilization and social progress, it also overwhelmed their nerves and depleted their material bodies. As Réveillé-Parise explained, contemporary scholars and *littérateurs* were no less driven than their predecessors by the "lively passion" for study; but they were now "devoured" by civic duties, too.[50] By participating in the social whirlwind, intellectuals multiplied the passions to which they were susceptible, and their ardent internal constitutions drove them to be just as consumed by public affairs as they were by the quest to create something glorious and immortal.

Thus, whereas Lorry and Tissot sought largely to correct the poor eating habits and obsessive work habits of *gens de lettres*, Brunaud, Réveillé-Parise, and Balzac glamorized those in this group. They portrayed great thinkers as burning with brain-centered vitality, a vitality that consumed itself and its physical trappings. In Balzac's most dramatic formulation of this idea, the truly dangerous modern stimulant was thought itself—which, in the words of Physidor, one of his many fictional doctor characters, "eats the body, absorbs and destroys it," vivifying and killing humanity all at once.[51]

[49] Physicians were not, of course, the only sources of Balzac's reflections on modern gastronomic excesses: his most explicit inspiration was Anthelme Brillat-Savarin's famous gastronomical treatise *Physiologie du goût* (1st ed., 1826).

[50] Réveillé-Parise, *Physiologie et hygiène*, vol. 1, 48 and vol. 2, 432.

[51] *Les Martyrs Ignorés*, 967–972.

Bibliography

Anon. "Hypochondriaque" (*passion ou affection*). In Denis Diderot and Jean le Rond d'Alembert, *Encyclopédie, ou dictionnaire raisonné des sciences, des arts et des métiers*. Paris: Chez Briasson [et al.], 1751–65, 17 vols. University of Chicago: ARTFL Encyclopédie Project, Spring 2016. http://encyclopedie.uchicago.edu, edited by Robert Morrissey and Glenn Roe. Vol. 8 (1765), 408–409. Accessed 18 January 2018.

Arnaud, Sabine. *On Hysteria: The Invention of a Medical Category Between 1670 and 1820*. Chicago: University of Chicago Press, 2016.

Balzac, Honoré de. "Avant-Propos" to *La Comédie humaine* [1842]. In *La Comédie humaine*, 12 vols. Paris: Gallimard, Editions de la Pléiade, 1976–1981. Vol. 1.

———. *Louis Lambert* [1832].

———. "Traité des excitants modernes" [1839]. In *La Comédie humaine*. Vol. 12.

———. "Les Martyrs Ignorés" [1847]. In *La Comédie des Ténèbres*. Paris: Omnibus, 2006.

Berrios, German E. "Hypochondriasis: A History of the Concept." In *Hypochondriasis: Modern Perspectives on an Ancient Malady*, edited by Vladan Starcevic and Don R. Lipsitt, 3–20. New York: Oxford University Press, 2001.

———. "In the Name of Hygieia and Hippocrates: A Quest for the Preservation of Health and Virtue." In *New Medical Challenges during the Scottish Enlightenment*, edited by Guenter B. Risse, vol. 76, 135–169, Clio Medica. Amsterdam: Editions Rodopi, 2005.

Bonnet, Charles. *Essai analytique sur les facultés de l'âme* [1760].

Brillat-Savarin, Anthelme. *Physiologie du goût* [first ed. 1826].

Brockliss, Laurence and Colin Jones, *The Medical World of Early Modern France*. Oxford: Clarendon, 1997.

Broussais, François-Joseph-Victor. *De l'irritation et de la folie*. [1828–39]. Paris: Fayard, 1986.

Brunaud, Étienne. *De l'hygiène des gens de lettres ou Essai médico-philosophique sur les moyens les plus propres à développer ses talents et son aptitude naturelles pour les sciences, sans nuire à sa santé et sans contracter des maladies*. Paris: Méquignon, 1819.

Buffon, Georges Louis Leclerc, comte de and Louis-Jean-Marie Daubenton. *Histoire naturelle générale et particulière: avec la description du Cabinet du Roy*. Paris: Imprimerie royale, 1749–87. 21 vols.

Cabanis, Pierre-Jean-Georges. *Rapports du physique et du moral de l'homme* [1802]. Paris: Slatkine, 1980.

———. *On the Relations Between the Physical and Moral Aspects of Man*. Edited by George Mora and Translated by Margaret Duggan Mora, 2 vols. Baltimore: Johns Hopkins University Press, 1981.

Chalmin, Ronan and Anne Vila. "Malade de son génie': raconter les pathologies des gens de lettres, de Tissot à Balzac." *Dix-huitième siècle* 47 (2015): 55–71.

Charpentier, C. A. T. *Essai sur la mélancolie*. Paris: Farge, 1803.

Charrier-Vozel, Marianne. "Sociabilités de la maladie: des manuels épistolaires aux lettres de Mme Riccoboni, de Mme du Deffand, de Mme d'Épinay, et de Mlle de Lespinasse." *Dix-huitième siècle* 4, no. 1 (2015): 231–243.

Condillac, Etienne Bonnot de. *Traité des sensations* [1754].

Cottom, Daniel. *Cannibals and Philosophers: Bodies of Enlightenment*. Baltimore: Johns Hopkins University Press, 2001.

Diderot, Denis. *Éléments de physiologie* [1778]. In *Œuvres complètes*, edited by Herbert Dieckmann, Jacques Proust, Jean Varloot, et al., 34 vols, anticipated. Paris: Hermann, 1975–. Abbreviated here as DPV. Vol. 17.

———. *Lettre sur les sourds et muets*. In DPV, vol. 4.

Edelman, Nicole. *Les Métamorphoses de l'hystérie: Du début du XIXe siècle à la Grande Guerre*. Paris: Editions de la Découverte, 2003.

Emch-Dériaz, Antoinette. "The Non-Naturals Made Easy." In *The Popularization of Medicine, 1650–1850*, edited by Roy Porter, 134–159. London and New York: Routledge, 1992a.

———. *Tissot, Physician of the Enlightenment*. New York: Peter Lang, 1992b.

Goldstein, Jan. *Console and Classify. The French Psychiatric Profession in the Nineteenth Century*. Chicago: Chicago University Press, 1987.

Hagner, Michael. "Skulls, Brains and Memorial Culture: On Cerebral Biographies of Scientists in the Nineteenth Century." *Science in Context* 16, nos. 1/2 (2003): 195–218.

Hatfield, Gary. "Remaking the Science of Mind: Psychology as Natural Science." In *Inventing Human Science: Eighteenth-Century Domains*, edited by Christopher Fox, Roy Porter, and Robert Wokler, 184–231. Berkeley: University of California Press, 1995.

Hecquet, Philippe. *La Médecine, la chirurgie et la pharmacie des pauvres* [1740].

Kassouf, Susan. "The Shared Pain of the Golden Vein: The Discursive Proximity of Jewish and Scholarly Diseases in the Late Eighteenth Century." *Eighteenth-Century Studies* 32, no. 1 (1998): 101–110.

Le Camus, Antoine. *La Médecine de l'esprit* [1753].

Le Yaouanc, Moïse. *Nosographie de l'humanité balzacienne*. Paris: Librairie Maloine, 1959.

Lorry, Charles. *Essai sur l'usage des alimens, pour servir de commentaire aux livres diététiques d'Hippocrate* [1754], 2nd ed., 2 vols. Paris: Vincent, 1757.

Lotterie, Florence. *Le Genre des Lumières, Femme et philosophe au XVIIIe siècle*. Paris: Classiques Garnier, 2013.

Louyer-Villermay, Jean-Baptiste. *Recherches historiques et médicales sur l'hypocondrie, isolée, par l'observation et l'analyse, de l'hystérie et de la mélancolie*. Paris: Méquignon, 1802.

———. *Traité des maladies nerveuses ou vapeurs, et particulièrement de l'hystérie et de l'hypocondrie*, 2 vols. Paris: Méquignon, 1816.

Marquer, Bertrand. "De l'épigastre au ventre: œconomia animale et économie du corps social." *Romantisme* 154, no. 4 (2011): 53–64.

———. "Portrait de l'artiste en dyspeptique." In *La Cuisine de l'œuvre au XIXe siècle: Regards d'artistes et d'écrivains*, edited by Bertrand Marquer and Éléonore Reverzy, 63–75. Strasbourg: Presses Universitaires de Strasbourg, 2013.

Micale, Mark S. *Hysterical Men: The Hidden History of Male Nervous Illness*. Cambridge, MA: Harvard University Press, 2008.

Nicoli, Miriam. *Les savants et les livres. Autour d'Albrecht von Haller (1708–1777) et Samuel-Auguste Tissot (1728–1797)*. Geneva: Slatkine, 2013.

Pasco, Alan. *Sick Heroes: French Society and Literature in the Romantic Age 1750–1850*. Exeter: University of Exeter Press, 1997.

Petetin, Jacques-Henri-Désiré. *Mémoire sur la découverte des phénomènes que prèsentent la catalepsie et le somnabulisme* [1787].

Pigeaud, Jackie. *Aux portes de la psychiatrie: Pinel, l'ancien et le moderne*. Paris: Aubier, 2001.

Pilloud, Séverine. *Les mots du corps: Expérience de la maladie dans les lettres de patients à un médecin du 18e siècle: Samuel Auguste Tissot*. Lausanne: Éditions BHMS, 2013.

Pinel, Philippe. *Nosographie philosophique, ou La méthode de l'analyse appliquée à la médecine*, 2 vols. Paris: Maradan, 1797.

Ramazzini, Bernardino. *De Morbis Artificium diatriba* [1700]. Translated as *Treatise of the Diseases of Tradesmen, Shewing the Various Influence of Particular Trades Upon the State of Health*. London: Andrew Bell, Ralph Smith, Daniel Midwinter et al., 1705.

Réveillé-Parise, Joseph-Henri. *Physiologie et hygiène des hommes livrés aux travaux de l'esprit ou Recherches sur le physique et le moral, les habitudes, les maladies et le régime des gens de lettres* [1834], 2 vols. Paris: Chez G.-A. Dentu, 1843.

Rey, Roselyne. *Naissance et développement du vitalisme en France de la deuxième moitié du dix-huitième siècle à la fin du Premier Empire*, SVEC 381. Oxford: Voltaire Foundation, 2000.

Ribard, Dinah. "Pathologies intellectuelles et littérarisation de la médecine: Une voie pour l'histoire du travail intellectuel." In *Littérature et médecine: Approches et perspectives (XVIe–XIXe siècle)*, edited by Andrea Carlino and Alexandre Wenger, 113–134. Geneva: Droz, 2007.

Roman, Myriam. "Avatars romanesques du penseur chez Mme de Staël, Balzac et Hugo." *Romantisme* 34, no. 124 (2004): 89–102.

Rousseau, George S. "Introduction." In *Hypochondriasis: A Practical Treatise, 1766*, edited by John Hill, i–xii. Los Angeles: William Andrews Clark Memorial Library, 1969.

Rousseau, Jean-Jacques. "Préface à Narcisse ou l'Amant de lui-même" [1752].

Roussel, Pierre. *Système physique et moral de la femme* [1775].

Spang, Rebecca L. *The Invention of the Restaurant: Paris and Modern Gastronomic Culture*. Cambridge, MA: Harvard University Press, 2000.

Spary, E. C. *Eating the Enlightenment: Food and the Sciences in Paris, 1670–1760*. Chicago: University of Chicago Press, 2012.

Tissot, Samuel-Auguste. *De la santé des gens de lettres*, 3rd rev. ed. Lausanne: Grasset, 1775.

———. *An Essay on Diseases Incident to Literary and Sedentary Persons, with Proper Rules for Preventing Their Fatal Consequences, and Instructions for Their Cure*. Translated by James Kirkpatrick. London: Norse and Dilly, 1769.

Trillat, Étienne. *Histoire de l'hystérie*. Paris: Éditions Seghers, 1986.

Vidal, Fernando. *Les sciences de l'âme, XVI–XVIIIe siècle*. Paris: Honoré Champion, 2006.

Vila, Anne C. "The Philosophe's Stomach: Hedonism, Hypochondria, and the 'New' Intellectual in Enlightenment France." In *Cultures of the Abdomen: Dietetics, Digestion, and Fat in the Modern World*, edited by Christopher E. Forth and Ana Cardin-Coyne, 89–104. New York: Palgrave, 2005.

Voltaire [Françcois-Marie Arouet]. *Œuvres complètes de Voltaire. Correspondance générale*, 66 vols. Paris: Chez Antoine-Augustin Renouard, 1819–1825. Vol. 56.

———. "Sensation," *Dictionnaire philosophique* (1764). In *Oeuvres complètes de Voltaire / The Complete Works of Voltaire*, 143 vols. Geneva: Institut et Musée Voltaire; Toronto: University of Toronto Press; Oxford: Voltaire Foundation, 1968. Vol. 36.

———. "Passions," *Questions sur l'Encyclopédie* (1774). In *Oeuvres complètes de Voltaire*, vol. 42B.

Williams, Elizabeth A. *The Physical and the Moral: Anthropology, Physiology and Philosophical Medicine in France, 1750–1850*. Cambridge: Cambridge University Press, 1994.

———. "Stomach and Psyche: Eating, Digestion, and Mental Illness in the Medicine of Philippe Pinel." *Bulletin of the History of Medicine* 84 (2010): 358–386.

Zimmermann, Johann-Georg. *Von der Erfahrung in der Arzneikunst* [1763].

CHAPTER 8

Digestion and Brain Work in Zola and Huysmans

Manon Mathias

This essay examines links between digestion and brain work in the figure of the writer in nineteenth-century France through readings of health advice books and novels. Health manuals from this period place a notable emphasis on diet and the stomach for overall health, but by focusing on some of these texts alongside novels centred on the viscera, the essay argues that the interest in digestion in both forms of writing is a means of elevating the mind.

Diet advice has been a regular feature of social life as far back as the time of Hippocrates.[1] But France led the way in the publication of "hygiene treatises" from the 1800s, and by the mid-nineteenth century, a veritable "avalanche" of such texts was being published.[2] Optimism regarding the power of science and the attainability of health; dismay over the effects of urban life; and the spread of public literacy assured

[1] See Temkin, *Hippocrates in a World of Pagans and Christians*, 45–47.

[2] Smith, *Clean: A History of Personal Hygiene and Purity*, 301.

M. Mathias (✉)
University of Glasgow, Glasgow, UK

M. Mathias and A. M. Moore (eds.), *Gut Feeling and Digestive Health in Nineteenth-Century Literature, History and Culture*, Palgrave Studies in Literature, Science and Medicine,
https://doi.org/10.1007/978-3-030-01857-3_8

an ever-expanding market for these books throughout the century.[3] Yet whereas France's leading role in the development of public health is well documented,[4] the importance of individual hygiene has received much less attention.

Digestive Hygiene

In his influential *Traité d'hygiène publique et privée* (*Treatise of Public and Private Hygiene*, 1844), the Strasbourg physician and hygienist, Michel Lévy, defined hygiene as "the Art of conserving health".[5] Although today "hygiene" is used interchangeably with "cleanliness", up until the end of the nineteenth century the term referred to "precepts or rules to follow to avoid illness or the causes that can lead to it".[6] It is this earlier sense of the word that is used in this essay, which examines the importance of the digestive system for the identity of the scholar.

The importance of food and eating in identity building is an area of research that has received recent attention in sociological, anthropological and cultural studies. Peter Scholliers proposes a "close relationship between identification and food",[7] for example, and Deborah Lupton argues that "food habits and preferences are central practices of the self".[8] Such studies, however, do not go beyond the ingestion of food and rarely concentrate on digestion. Historians Christopher Forth and Ana Carden-Coyne, in their introduction to essays on the belly from the Enlightenment to the present, also point out that "the abdomen remains an integral aspect of our sense of self as well as a favoured site of self-fashioning",[9] but the essays in their volume mostly examine food in the sense of cooking practices, eating disorders and nutrition, with digestion remaining marginal.

[3] See Whorton, *Inner Hygiene*, 37–38.

[4] See La Berge, *Mission and Method*.

[5] "l'Art de conserver la santé" (*Traité d'hygiène publique et privée*, I, 45).

[6] "les préceptes hygiéniques à suivre pour éviter ces maladies ou les causes qui peuvent les provoquer" (Barthés, Considérations sur l'hygiène des gens de lettres, 20). See also Whorton, *Inner Hygiene*, 31–32.

[7] *Food, Drink and Identity*, 9.

[8] *Food, the Body and the Self*, 15.

[9] *Cultures of the Abdomen*, 2.

Michel Foucault considered regimen and dietetics as examples of "technologies of the self", that is, ways in which an individual "acts upon himself".[10] Foucault's focus was on ancient Greek culture and the Roman Empire, when regimen was "a fundamental category through which human behaviour could be conceptualized".[11] According to Foucault, this focus on taking care of oneself was overshadowed in the modern age by the Delphic principle of "knowing oneself".[12] Indeed, with the growing focus in the eighteenth and nineteenth centuries on advancing knowledge and human progress, the thinking self was largely privileged at the expense of a holistic view of the individual which incorporates the body, and especially "the lower body".[13]

Yet, there was at the same time an increasing preoccupation with the viscera and their importance for well-being. French physician Réveillé-Parise insisted in 1834, for example, that "he who digests badly cannot hope for stable health or a long life".[14] There was also a particular interest in food and eating in relation to the scholar, as the stomach became a major centre of concern for the thinking individual. As Anne Vila has shown, in medical treatises in the second half of the eighteenth century, a picture emerged of the scholar as "a dysfunctional type whose range of potential constitutional ailments is symptomatically mapped out with an almost exclusive focus on two centres of pathological activity: the head and the stomach".[15] Such observations not only appeared in medical texts but also in writers' own views of themselves, as thinkers such as Voltaire, Rousseau, Kant and the Marquis de Sade placed food and eating at the centre of their philosophical reflections.[16] Sade, for instance, inverted Enlightenment medicine by portraying gluttony as a principled, "'philosophical' virtue",[17] and Rousseau used reflections on diet

[10] "The Technologies of the Self," 19.

[11] *The Use of Pleasure*, 101. See also Foucault, *The Care of the Self*, 100.

[12] "The Technologies of the Self," 19–22.

[13] Stallybrass and White, *The Politics and Poetics of Transgression*, 4.

[14] "Quiconque digère mal ne peut espérer ni une santé stable, ni une longue vie" (*Physiologie et hygiène des hommes adonnés aux travaux de l'esprit*, I, 166).

[15] "Enlightened Minds and Scholarly Bodies," 210–211.

[16] See Moore, Alison. "Kakao and Kaka. Chocolate and the Excretory Imagination in Nineteenth-Century Europe"; Mininger and Peck, "Kantstipation"; and Spary, *Eating the Enlightenment*.

[17] Vila, "Enlightened Minds and Scholarly Bodies," 217.

to examine the deleterious effects of civilization, promoting open-air, "simple" eating practices instead.[18]

It was also in the later eighteenth century that health manuals started to be published in France specifically targeting the "homme de lettres", paying considerable attention to diet. The text that had the greatest impact in this field was the bestselling *De la santé des gens des lettres* (*The Health of the Man of Letters*, 1768) by the Swiss physician, Samuel-Auguste Tissot, which suggested that scholars needed medical supervision to counteract the effects of their learning.[19] Such preoccupations gained significant traction in the nineteenth century as the stomach occupied an increasingly central place in French culture. It was in this century that the restaurant became a popular form of commercial endeavour; that gastronomy developed as a coded set of practices; and that the digestive system came to be placed at the centre of new medical disciplines such as pathological anatomy and psychosomatic medicine.[20] French physician François Broussais, for example, developed a gastro-centric pathology in the early decades of the century, and Philippe Pinel foregrounded the stomach in his influential investigations of mental illness.[21] Health manuals or "hygiene treatises" also now started to proliferate beyond the upper classes, targeting a range of different types of workers, and these placed a notable emphasis on digestion. In his influential *Treatise of Public and Private Hygiene*, for example, Michel Lévy claimed that care of one's digestive system was one of the most powerful means of influencing both mental and physical health.

The main focus in the hygiene treatises of the nineteenth century was on the means of preventing digestive ailments through daily practices. Tissot had provided some tips on healthy living in his *Health of the Man of Letters*, but in the nineteenth-century manuals the practical and preventative approach to health was at the centre of concern. Réveillé-Parise, for example, stressed his aim of writing a "useful",

[18] See Onfray, "Rousseau ou la voie lactée"; Spang, *The Invention of the Restaurant*, 34–63.

[19] Vila, "The Philosophe's Stomach," 94.

[20] See Spang, *The Invention of the Restaurant*; Ferguson, *Accounting for Taste*; and Miller, *A Modern History of the Stomach*.

[21] See, Ackerknecht, "Broussais or a Forgotten Medical Revolution"; Elizabeth Williams, "Stomach and Psyche."

"practical" book of "everyday practices",[22] and Strasbourg physician Étienne Brunaud distanced himself from Tissot by asserting that his own book did not deal with the knowledge and treatment of illness but rather with "the means of *preserving* men of letters from the large number of ailments which are the result of the relentless work schedule to which they are devoted".[23] This shift in emphasis was reflected in the titles of the books which now explicitly referenced "hygiene". Michel Lévy even went so far as to underline the difference between medicine—dealing with "illness, its causes, effects and remedies"—and hygiene, focused on "health, its conditions and its guarantees".[24]

The methods advocated in the hygiene manuals to preserve health were based on the principle of harmony or equilibrium between different parts of the body and especially between body and mind. Paul Gaubert, physician and long-term sufferer of gastric ailments, referred in *Hygiène de la digestion* (*Digestive Hygiene*, 1845) to "the harmony of functions", for example, as a form of "perfection" towards which "man must always strive",[25] and Brunaud argued that "there exists between the brain and the nutritive system a reciprocal dependence which cannot be ignored".[26] Repeated references to harmony and equilibrium demonstrate the aim of the hygienic texts to redress lack of balance through practical measures.[27]

All the health manuals I have examined, however, point to the ways in which such harmony is disrupted through mental activity. Brunaud noted that prolonged mental exertion "destroys the harmony between the functions of the cerebral organ and those belonging to the nutritive

[22] "un livre véritablement utile," "[cet] ouvrage a été fait pour être utile d'une manière tout-à-fait pratique et presque journalière" (*Physiologie*, I, iv, v).

[23] "des moyens de *préserver* les gens de lettres du grand nombre de maladies qui sont le résultat des travaux opiniâtres auxquels ils se livrent" (*De l'Hygiène des gens de lettres*, x, added emphasis).

[24] "la maladie avec ses causes, ses effets et ses remèdes," "la santé avec ses conditions et ses garanties" (*Traité*, 49).

[25] "l'harmonie des fonctions" "le type d'une perfection [...] vers laquelle [l'homme] doit toujours tendre" (*Hygiène de la digestion*, 121–122).

[26] "Il existe entre [...] le cerveau et l'ensemble de la vie nutritive une dépendance réciproque qu'on ne peut méconnaître" (*De l'Hygiène des gens de lettres*, 27–28).

[27] See Réveillé-Parise, *Physiologie*, I, 317, 346, 369, 371; Gaubert, *Hygiène de la digestion* 13; and Barthés, *Considérations*, 7.

system".[28] Réveillé-Parise argued that "a weakening of the digestive system is a particular trait amongst illustrious men",[29] and suggested that digestive health was in inverse relation to intellect.[30] Author and politician Emile Deschanel also pondered in his *Physiologie des écrivains et des artistes* (*Physiology of Writers and Artists*, 1864) whether physical health and mental health might be at odds with genius.[31] Bad digestion thus became a central element in the identity of the man of letters.

The ailment most often associated with literary life in the health manuals was constipation,[32] considered the most perilous ailment of civilized society by the late eighteenth century.[33] The anxiety surrounding this complaint was part of an increasing preoccupation in the literature, medicine and thought of the period with the effects of human progress (scientific, technological, intellectual) on health and morality. In particular, there was an increasing emphasis on the nefarious effects of a sedentary and cerebral lifestyle in contrast to the idealized active life of the country farmer.[34] The link between (in)digestion and mental capacities was thus part of a wider question on the relationship between civilization and health.

Whereas some showed alarm regarding the impact of deskbound living on the body, however, the health manuals published in France at this point foregrounded scholarly indigestion in ways which validated and even venerated this condition. Brunaud provided a long list of illustrious thinkers suffering from bad digestion, for example,[35] and noted "what a price [scholars] must pay for the great and glorious prerogative of enlightening the human race!"[36] He fetishized the thinker's condition

[28] "détruit peu à peu l'harmonie qui existe entre les fonctions de l'organe cérébral, et celles qui appartiennent au système général de la vie nutritive" (*De l'Hygiène des gens de lettres*, 17).

[29] "Aussi l'affaiblissement du système digestif semble-t-il particulier aux hommes illustres" (*Physiologie*, I, 35).

[30] See *Physiologie*, II, 240.

[31] *Physiologie des écrivains et des artistes*, 158.

[32] See Lévy, *Traité*, 201, 261, 736–737; Gaubert, *Hygiène de la digestion*, 172; and Réveillé-Parise, *Physiologie*, I, 107, II, 49, 291.

[33] See Whorton, *Inner Hygiene*, xi–xii, 33.

[34] See Réveillé-Parise, *Physiologie*, II, 105; Gaubert, *Hygiène de la digestion*, 241, 245.

[35] See *De l'Hygiène des gens de lettres*, 60–69.

[36] "qu'ils payent cher alors la belle et glorieuses prérogative d'éclairer le genre humain!" (*De l'Hygiène des gens de lettres*, 74).

by referring to "the exquisite frailty and sensitivity of the scholar's stomach".[37] Lévy similarly valorized the connection between bad digestion and intellect, noting that an intense connection between the viscera and the brain generally existed "in those of superior intelligence".[38] Réveillé-Parise even suggested that digestive breakdown raised the individual "beyond other mortal beings",[39] and he noted that some claimed to evaluate a man's genius based on the state of his stomach.[40] Whilst highlighting the physical dangers of the scholarly lifestyle, the hygiene texts presented a strong link between digestive problems and genius in ways which elevated gastric disorders.

Hygiene and the Novel: Zola

Given the imbrication of medical discourse and French literature in this period, it would be surprising not to find references to digestive hygiene in novels from this time. Indeed, in Balzac's *La Peau de chagrin* (1831), Raphaël's decline is largely due to "excessive intellectual work"[41] manifested in laborious digestion. George Sand, on the other hand, presents positive images of the "paysan", who, in his respect for natural rhythms and reflective eating, ensures a well-balanced existence.[42] The constant presence of food in general in the nineteenth-century French novel has been noted by critics[43] but the issue of digestion has not been analysed, and food is limited to its sociological connotations or its status as an indicator of reality.[44] Two authors who stand out in relation to digestive hygiene

[37] "la faiblesse et la sensibilité exquise de l'estomac" (*De l'Hygiène des gens de lettres*, 72). See also 53.

[38] "accompagne ordinairement les intelligences supérieures" (*Traité*, 80).

[39] "au-dessus du reste des mortels" (*Physiologie*, 102).

[40] "de nos jours, on a même été jusqu'à prétendre évaluer le génie d'après l'état de l'estomac" (*Physiologie*, II, 35).

[41] "des excès de pensée" (Balzac, *La Peau de chagrin*, 288).

[42] For more on food in Sand and Balzac, see Brown, *Fictional Meals*, 23–43, 55–89.

[43] See, for example, Jean-Pierre Richard's renowned statement, "on mange beaucoup dans les romans de Flaubert" (*Stendhal, Flaubert*, 137).

[44] See, for example: Brown, *Fictional Meals*, 6; Ruth Cruickshank, "Eating, Drinking and Re-Thinking: Marguerite Duras's *Moderato Cantabile* (1958)," 3, 11; and Auguste Dezalay, "Ceci dira cela: Remarques sur les antécédents du *Ventre de Paris*," 34–35. Sonnenfeld sees food as wholly ideological for Zola (Sonnenfeld, "Emile Zola: Food and Ideology," 606).

are Zola and Huysmans, prime examples of the novelistic preoccupation with the body, especially the lower body. This is particularly true of Zola's *Le Ventre de Paris* (*The Belly of Paris*, 1873). In his planning notes for this novel, Zola writes that "the stomach dominates the action" and "the general idea is: the stomach".[45] But the novel insists on this topic only in order to reject it, privileging instead the life of the mind.

Food is associated with dirt and putrefaction in Zola's *Belly of Paris*. The smell of fish in the market, for instance, is repeatedly linked with decay—"the stench rose and the air grew heavy with muggy pestilential exhalations" (200)—and this stink of rotting flesh "intermingles with the smell of mud from neighbouring streets" (200).[46] This "mud" was in fact faeces, as the streets would have been covered with the overflow from dung heaps. Despite the monumental sewer system developed under Baron Haussmann (the Prefect of the Seine) in the 1850s and 60s, these were primarily storm sewers designed to solve the city's drainage problems and clean up the Seine: they were not designed for human waste disposal. Cesspools thus continued to be in common use in Paris through the third quarter of the century, and excrement was not flushed into the sewers until 1895.[47]

The convergence of food and excrement recurs in Zola's novel when the protagonist, Florent, feels overwhelmed by the smell of sour cabbages, rotten apples and vegetables thrown into a pile of dung (387). Food is associated with excrement, decomposition and disease. It is even regarded as a form of death. The food market, Les Halles, for example, appears to Florent as "a vast ossuary, a place of death, littered with cadavers, a charnel house reeking with foul smells and putrefaction" (299).[48] Death and decay are also at the centre of the renowned description of the cheeses: their melting moulds are akin to "gaping wounds" (333)[49]

[45] "le ventre domine l'action" (Zola, *Le Ventre de Paris*, 19); "l'idée générale est: l'estomac" (Zola, *Le Ventre de Paris*, 449).

[46] "la puanteur monta, alourdit l'air d'une buée pestilentielle". "des senteurs de chairs tournées se mêlèrent aux soufflés fades de boue qui venaient des rues voisines."

[47] See La Berge, *Mission and Method*, 216; Monestier, *Histoire et bizarreries sociales des excréments*, 124–125; and Gandy, 30–31.

[48] "un vaste ossuaire, un lieu de mort où ne traînait que le cadavre des êtres, un charnier de puanteur et de décomposition."

[49] "des blessures mal fermées."

exuding "bitter fermentations" (332)[50] which culminate in "a cacophony of foul smells" (340).[51] The cheeses are personified as diseased, dying or dead: afflicted by "a shameful illness" (333),[52] they emit "bitter cries like those of a death agony" (335)[53] and are compared with skulls and rotting corpses (332, 333).

Furthering the connection between food and putrefaction in *The Belly of Paris* is the portrayal of digestion, a process that is devalorized through its association with bourgeois complacency and conservatism. The exemplary middle-class family, the Quenus, for example, live in "heavy, well-fed prosperity" (100),[54] and Florent is overpowered by "the blissful complacency" of their "continuous digestion and physical flourishing" (154).[55] The marketplace itself, les Halles, is also "a gratified beast constantly digesting" (205).[56] The process of digestion is linked with bourgeois self-righteousness and self-gratification. Ultimately, a solid, regularly used digestive system is associated with death in this text. The fat people ("les Gras") are in the end not only "sweating health" (422)[57] but "bursting with health, welcoming another day of gorging and digesting" (422).[58] Zola uses the word "crever" here for "to burst" which is also an informal term for dying. In the final scene, Lisa Quenu, the symbol of corpulent bourgeois self-righteousness, stands "lifeless" (423) and "motionless" (424).[59] The greasy health embodied in this character is a form of death.

If the digestion of food is associated with filth and decay on a literal and moral level, the inability to digest is inversely elevated. This is particularly the case for Florent, one of the few characters to show strong moral values. Florent can barely eat and is constantly overcome by

[50] "ces fermentations âcres."

[51] "une cacophonie de soufflés infects."

[52] "une maladie honteuse."

[53] "des râles de limbourg arrivèrent [...] aigres et amers, comme soufflés par des gorges de mourants."

[54] "prospérité alourdie."

[55] "la lâcheté heureuse de cette digestion continue du milieu gras où il vivait."

[56] "la bête satisfaite et digérant."

[57] "la charcuterie suait de nouveau la santé, une santé grasse."

[58] "crevant de santé, saluant un nouveau jour de belle digestion."

[59] "sans vie" "immobile."

faintness, nausea and stomach pains.[60] In keeping with the hygiene texts, he possesses the nervous disposition of the man of letters, and unlike Lisa, who is "nerve-free" (15) and "not nervous" (308), Florent is prone to severe nervous anxiety (201, 204). Some critics see him as a saintly character, with his hunger a form of martyrdom.[61] But Florent is also a scholar, referred to as "le professeur" (211, 345) and regularly engaged in thinking, dreaming and writing.

Florent's nightmarish visions are often caused by physical sensations such as those of being "lost in the superabundance of goods in the Halles", as argued by Zarifopol-Johnston.[62] But his nervous disposition is most frequently mentioned in those moments when he is engaged in thinking and writing. For example: "Tired from his days and his nerves all tight, [...] when he got home he did not go to bed, but stayed up and wrote" (313).[63] Florent stands for the scholar embracing the realm of thought rather than that of the body and its corrupted associations.[64]

There does seem to be a potential alternative between the gluttonous Quenu lifestyle and the painful existence of the scholar. Although Zola specialist David Baguley posits the opposition in this novel between the (negatively charged) materiality of food and the (valorized) "activities of the mind" as a gendered one,[65] a potentially positive model of physical well-being is offered in the portrayal of Madame François. This character rejects urban life and embodies rustic freshness and vitality. Whereas Lisa and the fishwives are compared with rotting meat, Madame François comes across as "a healthy and rugged plant that has grown like vegetables in the compost of a vegetable garden" (300).[66] In contrast to the overwhelming presence of death, disease and putrefaction in Paris and Les Halles, Madame François's domain in Nanterre is also filled with food but in a way that is suffused with life, fertility and nourishment:

[60] See Zola, *Le Ventre de Paris*, 101, 201, 243, 385.

[61] See Manzini, "Nutrition, Hunger and Fasting," 8, 24; Tunstall, "Crânement Beau Tout de Même," 183–184.

[62] Zarifopol-Johnston, "'Ceci Tuera Cela': The Cathedral in the Marketplace," 367.

[63] "Malheureux de sa journée, les nerfs tendus, [...] quand il rentrait, il ne se couchait pas, il écrivait." See also 384.

[64] For example, 277, 385.

[65] Baguley, "Le supplice de Florent," 94.

[66] "elle lui faisait l'effet d'une plante saine et robuste, grandie ainsi que les légumes dans le terreau d'un potager."

> the vegetable peelings [...] were now returned to the place where the vegetables had first sprung from the ground, to nourish new generations of cabbages, carrots and turnips. They would rise again as perfect produce, and return once more to spread along the footpath. (297)[67]

The restorative function of this space is highlighted in the repeated "re-"prefix ("elles repoussaient, elles retournaient") and in the image of Florent's own rebirth: "he was reborn in the pure air" (299).[68] Here, matter is not reviled as argued by Baguley,[69] but is rethought as a source of fertility and new life.

The countryside idyll, however, is a decidedly unreal, almost mythical space in Zola's novel. It is repeatedly linked with peace, purity and happiness (297–300), is hyperbolically viewed as "the health of the world" and is cloistered and hidden behind walls of hawthorns (298).[70] There is a magic quality to the food growing here, with a miraculously reappearing cabbage (297) and Florent's impression of hearing the vegetables growing (98). Madame François herself is introduced through her horse Balthazar, whose name suggests exotic tales rather than plodding reality, and she intervenes as a fairy godmother by rescuing Florent when others are happy to throw him into the gutter. She also enters Paris only at night like a figure in a fairy tale. Her status beyond the realm of positivist reality is confirmed in Claude's total inability to classify her: "Claude was quite embarrassed [...], He cast around for a while, then stammered: Madame François, Madame François—well, I don't know. I never thought about classifying her [...] She's neither Fat nor Thin, there!" (303).[71] Ultimately, Florent's dreamlike trip to Nanterre fades away, and the novel offers no viable alternative either to the queasy physicality of the Quenus or to Florent's digestive torment.

[67] "les épluchures des légumes [...] revenaient où les légumes avaient poussé, pour tenir chaud à d'autres générations de choux, de navets, de carottes. Elles repoussaient en fruits superbes, elles retournaient s'étaler sur le carreau."

[68] "il renaissait à l'air pur."

[69] "Le supplice de Florent," 94.

[70] "la santé du monde."

[71] "Claude fut très embarrassé [...] Il chercha, balbutia, 'mme François .... Mme François ... Non, je ne sais pas, je n'ai jamais songé à la classer ... Elle n'est ni dans les Gras ni dans les Maigres, parbleu!'"

At the moment when Florent is most deeply absorbed in his political plans, he not only rejects the processes of digestion but also becomes almost disembodied: "his feet seemed not to touch the ground, as if borne aloft by his desire to pursue justice" (356).[72] Like the health manuals which advise the individual to distance himself from the degrading reliance on the body,[73] Zola seems to foreground the digestive system as a means of denigrating the physical side of human existence and exalting the life of the mind.

Huysmans

This ideal of scholarly disembodiment is taken further by Huysmans. His novel of 1884, *A Rebours* (*Against Nature*), is often read as a warning against the risks of an unhealthy lifestyle cut off from the rest of society: nineteenth-century critics, for instance, read the text as a cautionary tale,[74] and Ludmilla Jordanova interprets the novel as a fictional "case study" demonstrating the dangers of immoderate regimen (112).[75] Indeed, the novel ends with the realization that "a new regimen" (310)[76] must be followed if the central character is to regain health and Raymond Prier considers physicality to triumph in this text.[77] Francesco Manzini, on the other hand, argues that nutrition and appetite are "trumped" in this novel by spirituality.[78] This essay reveals instead a constant tension in Huysman's novel between the foregrounding and denying of the body.

Against Nature seems to centre on the transcendence of bodily existence. Des Esseintes escapes to his country retreat as a way of rejecting contact with the city, and he famously travels in his mind rather than

[72] "il ne marchait plus à terre, comme soulevé par cette idée intense de se faire le justicier des maux qu'il avait vu souffrir."

[73] See Gaubert, *Hygiène de la digestion*, 223.

[74] See McGuinness, "Introduction," xiii.

[75] Jordanova, Ludmilla. "'A Slap in the Face of Old Mother Nature': Disease, Debility, and Decay in Huysmans's A Rebours." See also Segura, "The Dream of the Hermetic Utopia"; Jørgensen, "The Bacteriological Modernism of Joris-Karl Huysmans's *Against Nature*."

[76] "de nouvelles conditions d'hygiène."

[77] Prier, "Consumption to the Last Drop: Huysmans' Dyspeptic Tale of Eating, *A Rebours*," 196. Prier draws various links between digestion and aspects of *Against Nature*. For example, it is claimed of des Esseintes' dining room walls that "nothing could resemble more the texture of the intestinal wall" (187) and the porthole in this room is said to be "constipated" (187) by an aquarium. No explanation is offered for these comments.

[78] Manzini, "Nutrition, Hunger and Fasting," 31.

taking an actual trip to London. Des Esseintes' preferences for "bouillons" or meat essences also show his attempt to overcome the body, as this dish required a lengthy cooking process involving boiling things down to a purified essence.[79] It was a highly complex and artificial product requiring time-consuming and elaborate preparation, and in breaking down food, it liberated the body from its digestive duties. Partisans of this new cuisine contrasted the old-fashioned eater weighed down with "heavy, earthy foods", with a modern eater "released from corporeality by his subsistence on 'essences'".[80] Des Esseintes' predilection for such dishes shows his yearning for disembodiment, taken to its extreme when he removes the need to eat altogether and feeds himself through an enema which eliminates "the tedious and vulgar chore of eating" (304).[81]

Des Esseintes' preference for bouillons is also a comment on forms of writing. The link is suggested in Huysmans' preface where he refers to each chapter of the novel as a "coulis" or concentrate of a specialized subject (55). Des Esseintes himself admires a form of writing which he terms a "highly refined jus" (115).[82] His favourite writers are those such as Mallarmé whose work, he claims, results in "a condensed form of literature, a concentrated essence" (289).[83] Des Esseintes ponders the possibility of writing a novel "which concentrated into a few sentences the distilled juice of hundreds of pages", a novel in which every adjective would "open up such wide vistas that the reader could muse on its meaning […] for weeks on end, ascertaining the present, reconstructing the past, and divining the future […] in the light of this one unique epithet'" (291).[84] This form of literature is thus a reduction of the material

[79] See Spang, *The Invention of the Restaurant*, 1–2.

[80] Ibid., 48.

[81] "l'ennuyeuse et vulgaire corvée du repas."

[82] "aux sucs épurés."

[83] "une littérature condensée, un coulis essentiel."

[84] "un roman concentré en quelques phrases qui contiendraient le suc cohobé des centaines de pages […]. l'adjectif posé d'une si ingénieuse et d'une si définitive façon […] ouvrirait de telles perspectives que le lecteur pourrait rêver, pendant des semaines entières, sur son sens, tout à la fois précis et multiple, constaterait le présent, reconstruirait le passé, devinerait l'avenir des âmes des personnages, révélés par les lueurs de cette épithète unique."

to its purest essence beyond the earthly limitations of time and space. A novel written in this compressed style would lead to "a communion of thought" between writer and reader, "a spiritual collaboration"[85] that brings together "superior beings" scattered across the universe (291).[86] The writing leads to an untethering from the binds of earthly existence and a union of minds alone, a process which elevates the thinking person above other individuals.

Manzini argues that Des Esseintes is more concerned with fasting in a spiritual sense than with "scientistic discourses of nutrition" (31), but, through the prolonged metaphor of the bouillon, Huysmans was engaging with heated debates in contemporary biology and chemistry on the processes of digestion. Discussing prose poetry, for example, Des Esseintes sums up this form as "the osmazôme of literature" (291),[87] referring to the term for the savoury element of concentrated meat juice as used in Brillat-Savarin's gastronomic bestseller, *Physiologie du goût* (*Physiology of Taste*, 1825). This chemical term from the medico-culinary culture of the period might be an invitation for us to mock Des Esseintes if we remember its use in *Madame Bovary* (1857) by the ridiculed Monsieur Homais: "they were dazzled by his talk of aroma, osmazôme, juices and gelatines".[88] But the term also ties in with Des Esseintes' understanding of the literary process as a form of reduction, apparently beyond the confines of space and time, and he thus draws on the terminology of eating and digesting precisely in order to transcend the realm of the body, in a paradoxical co-dependence.

Such co-dependence is suggested in Huysmans' comments on painting in his other writings. In *En Hollande* (*In Holland*, 1877) and *Voyages aux cathédrales rouges* (*Visiting Red Cathedrals*, 1903), he alternates between the physiological processes of eating, digesting and excreting and ecstatic moments of transcendence. Alain Buisine has suggested that Huysmans' art criticism can be compared with the body's need to empty

[85] "une communion de pensée" and "une collaboration spirituelle."

[86] "dix personnes supéieures éparses dans l'univers."

[87] "l'osmazôme de la littérature."

[88] "il parlait arome, osmazôme, sucs et gélatine d'une façon à éblouir" (Flaubert, *Œuvres Complètes*, III, 235).

itself and loosen its bowels,[89] since it entails a defecation or flushing out of the real:

> painting as analysed and fantasised by [Huysmans] has no other function than to relieve [him of] reality. We might dare to define it in the most physiological of terms as a form of defecation, by reality itself, [...] of its own referential charge. The real discharges [itself] to become no more than painted surface.[90]

At the same time, however, Huysmans pays close attention to food and eating in the paintings that he views, thus providing a paradoxical contrast to his insistence on expunging food and materiality through art. Buisine suggests that Huysmans' art criticism is founded on "a double operation"[91] involving at once elevation and abasement, purification and contamination, sublimation and materialization. Whilst these comments relate to painting, Buisine's notion of doubling is valuable in considering the perception of literature in *Against Nature*, where the digestive system is present both as a constant reminder of the reality of life in the body and as a means of release from this reality.

Despite the aspiration towards transcendence, Huysmans' novel repeatedly foregrounds eating, digesting and excreting in ways which fuse allusions to food as excrement and excrement as food.[92] Horse droppings are "the colour of gingerbread" (215),[93] and excrement-infused water is compared with a bowl of coffee (208). Beef essence is a "muddy, salty liquid" left at the bottom of the pan (267),[94] in an image highly suggestive of faecal waste, not least due to the fact that the beef concentrate is produced by a device that breaks down food, acting as an

[89] Buisine, "Huysmans à fleur de peau," 344.

[90] "la peinture telle que l'analyse et surtout le fantasme l'auteur [..] n'aura d'autre fonction que de soulager le réel. Oserait-on la définir très physiologiquement comme une défécation par le réel lui-même [..] de sa propre charge référentielle? Le réel (se) décharge pour n'être plus que surface peinte."

[91] "une double opération" (Buisine, "Huysmans à fleur de peau," 347).

[92] For more on links between ingestion and excretion in Huysmans' other work, see Przybós, "Concerning a Physiological Geometry."

[93] "les pavés étaient enduits d'une boue de pain d'épice."

[94] "[le] jus bourbeux et salé, déposé au fond de la marmite."

artificial digestive system. Meat extract also reappears as one of the first meals taken by the protagonist Des Esseintes after moving off the enemas, when he is served a form of powdered meat "whose vague aroma of cocoa pleased his real mouth" (304).[95] The previous reference to beef concentrate prepares us to read this as potentially excremental, and the allusion to the "real" mouth draws attention to Des Esseintes' recent ingestion of sustenance through his other "mouth", the anus.

We might even say that the body "wins" in *Against Nature*, as the novel concludes with the submersion of the protagonist by dirty matter. Des Esseintes states that, "like a tidal wave, the waves of human mediocrity are rising up to the sky, and will engulf the refuge whose floodgates I unwillingly open" (318).[96] The necessity of opening the gates is a reference to the "health regimen" (310)[97] advocated by his doctor which requires des Esseintes to reconnect with the "filth", "mud", (311) and "slime" (318) of society.[98] His attempt at transcendence is in vain. Des Esseintes must face the filth as he realizes that he can "no longer deceive himself" (317).[99] In this tale of perversion and of going against nature, the protagonist ultimately moves away from deception and must let the filth in.

A similar immersion takes place for Florent in Zola's *Belly of Paris*, as "the filth from the greasy streets rose all around him" (415),[100] and Florent, the figure of the writer, is subsumed by the muck he tried to avoid. Such immersion of the individual by outer forces, and anxiety concerning this process, is posited by theorists such as psychologists, sociologists and anthropologists as the main motivation behind the barriers we as humans impose between clean and dirty, inside and outside, self and other. The invasion of filth is the ultimate threat to a stable sense of identity.[101] Yet whilst attempting to reject the physicality of the lowly body as a means of shoring up the scholar's identity, Zola and Huysmans also

[95] "dont le vague arôme de cacao plaisait à sa réelle bouche."

[96] "comme un raz de marée, les vagues de la médiocrité humaine montent jusqu'au ciel et elles vont engloutir le refuge dont j'ouvre, malgré moi, les digues."

[97] "conditions d'hygiène."

[98] "l'ordure," "la boue," "la bourbe," and "[la] fange."

[99] "il ne pouvait se le dissimuler."

[100] "autour de lui, montait la boue de ces rues grasses."

[101] See Kristeva, *Powers of Horror*, 3–4.

suggest an essential and potentially fruitful connection between writing and the digesting body.

The linking of food with filth ties Huysmans to his literary predecessors through a chain of intertextual echoes. Des Esseintes's "funeral feast" (83)[102] at the beginning of *Against Nature*, for example, notably features "smoked black pudding" (84)[103] which is a central element in Zola's *Belly of Paris* when Florent relates the story of his suffering as an escaped convict. Critic Jean Borie has pointed to certain foods in Zola's novel as key exemplars of its excremental tonality, especially blood pudding due to its fatty, sludgy texture.[104] The symbolic status of blood pudding is also suggested in Flaubert's *Bouvard et Pécuchet* (1881) where a discussion of digestive processes centres on "le boudin".[105] *The Belly of Paris* develops the merging of food and excrement in Florent's tale as he refers to the ingestion of filth when he was stuck "up to his stomach"[106] in mud: "with every effort he made to try to free himself the mud seemed to rise up towards his mouth" (150).[107] The image of the filth-filled mouth is reminiscent of a man's death in quicksand in Hugo's *Les Misérables* (1863), a novel which reverberates throughout *The Belly of Paris.*[108] In Hugo's novel, the man's mouth is filled with sand rather than mud,[109] but the passage acts as a preface to Valjean's ordeal in the sewers and prefigures the hero's descent into pestilential filth.

The intertwining of food and excrement operates not only on a thematic level for Zola but also in relation to the process of writing. In an analysis of *The Belly of Paris* as an allegory of Zola's authorial relationship with Hugo, critic Marie-Sophie Armstrong posits that the scene in the butcher shop where Florent tells his story is "a rewriting on a small scale of the Hugolian sewers".[110] The fact that digestion and excretion surface

[102] "repas de deuil."

[103] "des boudins fumés."

[104] Broie, "Les Fatalités du corps dans *Les Rougon-Macquart*," 1579.

[105] Flaubert, *Bouvard et Pécuchet*, 126.

[106] "jusqu'au ventre."

[107] "A chaque secousse qu'il tentait pour se dégager, la boue semblait monter à sa bouche."

[108] See Baguley, "Le supplice de Florent"; Zarifopolo-Johntson, "Ceci Tuera Cela"; and Dezalay "Ceci dira cela."

[109] Hugo, *Les Misérables*, 1318.

[110] Marie-Sophie Armstrong, "Hugo's 'égouts' and *Le Ventre de Paris*," 397.

in scenes where Zola is working through his anxious literary relationship with his predecessor suggests that the viscera are potentially significant for Zola's reflections on his status as a writer. Indeed, the episode with the blood pudding in *The Belly of Paris* is essentially engaged with the act of story-telling: Florent, in his position as teller of the tale, takes on the role of the author, and the passage moves between his tale of pain and starvation and the sickening process of making the blood pudding, in an atmosphere heavy with food and laborious digestion.[111] Zola's linking of the digestion/excretion pairing with the writing of literature also echoes Flaubert's famous assertion in 1855: "Shit rises in my mouth [...]. I want to make it into a paste with which to draw the 19th century".[112] The image of filth in the mouth is here again used in relation to writing, and digestive processes seem central for Zola and Flaubert's construction of their literary identity. The presence of blood pudding in *Against Nature* connects Huysmans and other major authors of his time in ways which foreground the intersection of digesting and writing.

Connections between digestion, excretion and literature come to the fore in Huysmans' *Against Nature* since, in this text, physicality and cerebrality are essentially interconnected. Like Florent, Des Esseintes possesses the typical nervous temperament of the thinking man. We repeatedly hear of his weak nervous system (78) and his brain is said to be overexcited by "neurosis" (194).[113] Although he does not write himself, Des Esseintes stands as a corollary for the author, and his intellectual stimulation alternates with eating and digesting. Des Esseintes moves, for example, between reading the paper and eating his "tartines" (103), between the library and the dining room (137), and between reflections on art and updates on his stomach (264). When he undertakes long spells of reading and particularly after reflections on literary criticism, he suffers from gut problems that flow from his head to his stomach (167).

Huysmans, however, does more than simply take up the connections between the digestive system and writing. Throughout *Against Nature*, literature itself is discussed in digestive terms. Villiers' work, for instance, is said to be "spasmodic" (284); Ambrosius is the author of "indigestible

[111] See *Le Ventre de Paris*, 148.

[112] "Il me monte de la merde à la bouche [...]. J'en veux faire une pâte dont je barbouillerais le XIXe siècle" (Flaubert, *Correspondance*, II, 600).

[113] "la névrose." See also 72, 99, 186.

homilies" (110)[114]; and Caesar's pithiness is a form of "unforgivable constipation" (103).[115] Such musings go beyond the standard metaphor of reading as consumption, as the writers' style itself is conceptualized as a form of (in)digestion. Rather than pitting cerebrality and art against the body, then, Huysmans reveals the former to be inherently intertwined with the latter. Even artifice, conceived in this text as a means of escaping nature in the sense of the material and the physical, is linked in the novel with food and digestion. Des Esseintes tries various "tricks" to "deceive" his "distrustful stomach" (267),[116] and in his creation of new bouillons, he shows his allegiance with artifice or fakery as he becomes "a fake gourmet" (305).[117] Even religion is shown to be rife with duplicity in the form of fake food (314).

The connection between art and the digesting body both enables a release from material reality through the process of writing and at the same time reminds the reader of the inevitability of life in the body. Rather than posing an either/or, Zola and Huysmans posit a both/and. For example, although Florent seems to be defeated by filth at the end of *Belly of Paris*, the novel concludes with the release of a bird, the ultimate symbol of freedom and elevation.

To conclude, the hygiene texts and the novels examined here use references to digestion as a way of foregrounding the mind and transcending the body. By drawing attention to digestive problems, the novelists and hygienists of nineteenth-century France set the scholar apart from the rest of society and the functions of digestion and putrefaction. In this way, the gut becomes a central element in the construction of authorial identity in the sense that bad digestion is promoted as a badge of honour. This identity, however, proves to be much more complex: rather than releasing themselves from the filthy body, Florent and Des Esseintes reveal instead an essential link between writing, digesting and excreting, and what we ultimately find is a paradoxical co-dependence between the body and the mind.

[114] "l'auteur d'indigestes homélies."

[115] "charnu, nourri, mais tourné à la graisse," "une constipation incroyable et indue."

[116] "des ruses qui réussissent à leurrer la défiance de l'estomac."

[117] "[un] faux gourmet."

Bibliography

Ackerknecht, Erwin H. "Broussais or a Forgotten Medical Revolution." *Bulletin of the History of Medicine* 27 (January 1953): 320–343.

Armstrong, Marie-Sophie. "Hugo's "égouts" and *Le Ventre de Paris*." *The French Review* 69, no. 3 (February 1996): 394–408.

Baguley, David. "Le supplice de Florent." *Europe* 46 (April–May 1968): 91–96.

Balzac, Honoré de. *La Peau de chagrin*. Paris: Flammarion, 1996.

Barthés, Jean-Pierre. *Considérations sur l'hygiène des gens de lettres*. Montpellier: J. Martel aîné, 1849.

Borie, Jean. "Les Fatalités du corps dans *Les Rougon-Macquart*." *Les Temps modernes* 24 (1968–1969): 1567–1591.

Brown, James W. *Fictional Meals and Their Function in the French Novel 1789–1848*. Toronto: Toronto University Press, 1984.

Buisine, Alain. "Huysmans à fleur de peau." *Nineteenth-Century French Studies* 17, nos 3–4 (1989): 342–354.

Carden-Coyne, Ana and Christopher E. Forth, eds. *Cultures of the Abdomen: Dietetics, Digestion and Obesity in the Modern World*. New York: Palgrave Macmillan, 2005.

Cruickshank, Ruth. "Eating, Drinking and Re-Thinking: Marguerite Duras's *Moderato Cantabile* (1958)." *Forum for Modern Language Studies* 49, no. 3 (2013): 300–312.

Dezalay, Auguste. "Ceci dira cela: Remarques sur les antécédents du *Ventre de Paris*." *Les Cahiers Naturalistes* 58 (1984): 33–42.

Ferguson, Priscilla Parkhurst. *Accounting for Taste: The Triumph of French Cuisine*. Chicago: Chicago University Press, 2004.

Foucault, Michel. "The Technologies of the Self." In *Technologies of the Self. A Seminar with Michel Foucault*, edited by L. Martin, H. Gutman, and Hutton, 16–49. London: Tavistock Publications, 1988.

———. *The Care of the Self. The History of Sexuality*, vol. 3. London: Penguin, 1990.

———. *The Use of Pleasure. The History of Sexuality*, vol. 2. Harmondsworth: Penguin, 1992.

Flaubert, Gustave. *Correspondance*. Edited by Jean Bruneau, 6 vols. Paris: Gallimard, 1973–2007.

———. *Œuvres Complètes*. Edited by Claudine Gothot-Mersch et al., 3 vols. Paris: Gallimard, 2001–2013.

———. *Bouvard et Pécuchet*. Edited by Stéphanie Dord-Crouslé. Paris: Flammarion, 2011.

Gandy, Matthew. "The Paris Sewers and the Rationalization of Urban Space." *Transactions of the Institute of British Geographers* 24, no. 1 (1999): 23–44.

Gaubert, Paul. *Hygiène de la digestion*. Paris: Dentu, 1845.

Hugo, Victor. *Les Misérables*. Paris: Gallimard, 1976.

Huysmans, Joris-Karl. *À Rebours*. Paris: Gallimard, 1983.

Jordanova, Ludmilla. "'A Slap in the Face of Old Mother Nature': Disease, Debility, and Decay in Huysmans's *A Rebours*." *Literature and Medicine* 15, no. 1 (Spring 1996): 112–128.

Jørgensen, Jens Lohfert. "The Bacteriological Modernism of Joris-Karl Huysmans's *Against Nature*." *Literature and Medicine* 31, no. 1 (Spring 2013): 91–113.

Kristeva, Julia. *Powers of Horror. An Essay on Abjection*. Translated by Leon Roudie. New York: Columbia University Press, 1982.

La Berge, Anne. *Mission and Method: The Early Nineteenth-Century French Public Health Movement*. Cambridge: Cambridge University Press, 1992.

Lévy, Michel. *Traité d'hygiène publique et privée* [1845], 2 vols, 2nd ed. Paris: Baillière, 1850.

Lupton, Deborah. *Food, the Body and the Self*. London: Sage, 1996.

Manzini, Francesco. "Nutrition, Hunger and Fasting: Spiritual and Material Naturalism in Zola and Huysmans." *Forum for Modern Language Studies* 49 (2013): 20–32.

McGuinness, Patrick. "Introduction." In *Against Nature*, edited by Joris-Karl Huysmans and translated by Robert Baldick, xiii–xxxvi. London: Penguin, 2003.

Miller, Ian. *A Modern History of the Stomach: Gastric Illness, Medicine and British Society, 1800-1950*. London: Pickering and Chatto, 2011.

Mininger, J. D. and Jason Michael Peck. "Kantstipation." In *Consuming Culture in the Long Nineteenth Century: Narratives of Consumption, 1700–1900*, edited by Tamara S. Wagner and Narin Hassan, 209–222. Lanham, MD: Lexington Books, 2007.

Monestier, Martin. *Histoire et bizarreries sociales des excréments des origines à nos jours*. Paris: Le Cherche Midi, 2012.

Moore, Alison. "Kakao and Kaka. Chocolate and the Excretory Imagination in Nineteenth-Century Europe." In *Cultures of the Abdomen* (2005): 51–69.

Onfray, Michel. "Rousseau ou la voie lactée." *Le Ventre des philosophes. Critique de la raison diététique*, 43–59. Paris: Le Livre de poche, 1989.

Prier, Raymond Adolph. "Consumption to the Last Drop: Huysmans' Dyspeptic Tale of Eating, *A Rebours*." In *Disorderly Eaters. Texts in Self-Empowerment*, edited by Lilan R. Furst and Peter W. Graham, 185–197. University Park, PA: Pennsylvania University Press, 1992.

Przybós, Julia. "Concerning a Physiological Geometry: The Metaliterary Content of J.-K. Huysmans' 'Sac au Dos'." *L'Esprit Créateur* 25, no. 4 (Winter 1985): 105–114.

Réveillé-Parise, Joseph Henri. *Physiologie et hygiène des hommes livrés aux travaux de l'esprit* [1834], 4th ed. Paris: G.-A. Dentu, 1843.

Richard, Jean-Pierre. *Stendhal, Flaubert.* Paris: Seuil, 1970.
Scholliers, Peter, ed. *Food, Drink and Identity: Cooking, Eating and Drinking in Europe Since the Middle Ages.* Berg: New York, 2001.
Segura, Debra. "The Dream of the Hermetic Utopia: *A Rebours* as Allegory for the World After Germ Theory." *Discourse* 29, no. 1 (Winter 2007): 49–76.
Smith, Virginia. *Clean: A History of Personal Hygiene and Purity.* Oxford: Oxford University Press, 2007.
Sonnenfeld, Albert. "Emile Zola: Food and Ideology." *Nineteenth-Century French Studies* 19, no. 4 (Summer 1991): 600–611.
Spang, Rebecca L. *The Invention of the Restaurant: Paris and Modern Gastronomic Culture.* Cambridge, MA: Harvard University Press, 2000.
Spary, E. C. *Eating the Enlightenment. Food and the Sciences in Paris, 1670–1760.* Chicago: University of Chicago Press, 2012.
Stallybrass, Peter and Allon White. *The Politics and Poetics of Transgression.* Ithaca, NY: Cornell University Press, 1986.
Temkin, Oswei. *Hippocrates in a World of Pagans and Christians.* Baltimore: John Hopkins, 1991.
Tunstall, Kate. "Crânement Beau Tout de Même': Still Life and *Le Ventre de Paris.*" *French Studies* 58, no. 2 (2004): 177–187.
Vigarello, Georges. *Concepts of Cleanliness.* Cambridge: Cambridge University Press. 1988.
Vila, Anne C. "Enlightened Minds and Scholarly Bodies from Tissot to Sade." *Studies in Eighteenth-Century Culture* 23 (1993): 207–220.
———. "The Philosophe's Stomach. Hedonism, Hypochondria and the Intellectual in Enlightenment France." In *Cultures of the Abdomen* (2005): 89–104.
Whorton, James C. *Inner Hygiene: Constipation and the Pursuit of Health in Modern Society.* Oxford: Oxford University Press, 2000.
Williams, Elizabeth, "Stomach and Psyche: Eating, Digestion, and Mental Illness in the Medicine of Philippe Pinel." *Bulletin of the History of Medicine* 84, no. 3 (Fall 2010): 358–86.
Zarifopol-Johnston, Ilinca. "'Ceci Tuera Cela': The Cathedral in the Marketplace." *Nineteenth-Century French Studies* 17, nos. 3–4 (Spring–Summer 1989): 355–368.
Zola. *Le Ventre de Paris.* Paris: Gallimard, 2002.

CHAPTER 9

Textual (In)Digestions in Flaubert, Zola and Huysmans: Accumulation, Extraction, Regulation

Larry Duffy

The nineteenth century saw a revolution in the way in which knowledge was accumulated, organised, ordered and produced. Concurrently, there were concerted attempts to regulate and conceptualise the human body and its functions. One regulatory conceptualisation can be found in historian Georges Vigarello's notion of "redressement" ["righting," or "straightening"]. Vigarello identifies a nineteenth-century preoccupation with the rectification of the deformed body and argues that this preoccupation and practices associated with it found analogies in education, whereby the body of accumulated knowledge and its dissemination were reordered according to taxonomic prerogatives informing "a system of basic and precise movements integrated within a veritable programme" ["une systématique de mouvements élémentaires et précis s'intégrant

L. Duffy (✉)
University of Kent, Canterbury, UK

M. Mathias and A. M. Moore (eds.), *Gut Feeling and Digestive Health in Nineteenth-Century Literature, History and Culture*, Palgrave Studies in Literature, Science and Medicine,
https://doi.org/10.1007/978-3-030-01857-3_9

dans un véritable programme"].[1] Implicit is that knowledge and new ways of organising it—in accordance with what Philippe Hamon refers to as the "cybernetic" function of the body, itself analagous to a text or archive storing information and internalising discourses absorbed from its ambient *milieux*—were incorporated within the body and actualised in bodily practices.[2]

Such incorporation, which we might term discursive incorporation, was a key preoccupation and practice of nineteenth-century writers. Incorporation is the process of knowledge and discourse becoming embodied; the coming into being of institutional bodies; the internalisation of discourse within the body as site of the functioning of disciplinary power; and at a basic level, the productive absorption within literary texts of extraliterary discourse.[3] Another potentially productive way of understanding discursive incorporation is through a very specific realm of bodily absorption, that of digestion, itself a major preoccupation of novelists in whose work representation of the body, and incorporation—ingestion and shaping—of knowledge about it, are privileged. The verb "digérer" ["digest"] until the seventeenth century meant "mettre en ordre" ["put in order"].[4] Despite the later shift in meaning, digestion as a *mise en ordre*, or ordering, is a fitting subject for nineteenth-century cultural representation, offering a useful way of understanding literary representations of digestion itself, not least since the nineteenth century, as well as being a moment when digestion became a significant theme in French medicine, popular culture and literature, was also precisely a time of putting things in order, both in the discursive and physiological realms.

This chapter will argue that major French authors broadly categorisable as "naturalist" (Flaubert, Zola, Huysmans), who were acutely concerned with the body as well as with knowledge and the ways in which it is assimilated, arranged and disseminated, envisaged digestion in both physiological and discursive terms.

[1] Georges Vigarello, *Le corps redressé*, 152. This and all other translations are my own.

[2] Philippe Hamon, "Du Savoir dans le texte," 494.

[3] Larry Duffy, *Flaubert, Zola and the Incorporation of Disciplinary Knowledge*.

[4] Paul Robert, *Le Petit Robert*, 738.

Flaubert

Flaubert's most quoted pronouncement on digestion is probably that found in the *Dictionnaire des idées reçues* [*Dictionary of Received Ideas*], lampooning the commonplace that "toutes les maladies viennent de l'estomac" ["all illnesses come from the stomach"].[5] In his fiction, the stomach has an understated but strong presence. In *Madame Bovary*, explicit mentions of the stomach are minimal.[6] Two out of the three instances of the term "estomac" and the sole instance of the term "épigastre" ["epigastrium"] occur only in the context of Emma's death, where Homais the pharmacist, reports "'intolerable pains in the epigastrium'" ["'des douleurs intolérables à l'épigastre'"] (significantly, over lunch, "after the first morsels" ["après les premiers morceaux"], that is, once digestion has commenced).[7] In the *avant-texte*, however, there is a curious allusion to the stomach—still resonant in the published text—suggestive of analogies between the disruptive ingestion of matter and the assimilation of knowledge. The recurrence of Emma's "maladie nerveuse" ["nervous malady"] prompts "de longues conférences" ["long discussions"] between Charles Bovary and his mother, who agree that Emma should not read novels. Charles's mother even denounces Emma's bookseller as practising a "métier d'empoisonneur" ["poisoner's trade"] (*MB*, 192). There is no digestive content here, other than the notion that literature is a poison disrupting the female sensorium. However, previous drafts reveal that medical discourse asserting precisely the dangers of reading for mind and body underpins this episode, linked in turn to Emma's death by actual bodily poisoning.[8] The final copyist's manuscript reveals of the long discussions that "the apothecary was involved in them" ["l'apothicaire en était"] (C, f° 199). The apothecary, orderer, classifier, prescriber and proscriber of what is to be ingested, and of reading material loaned to Emma from his library—and a poisoner in

[5] Gustave Flaubert, *Œuvres complètes*, II, 308.

[6] It has nevertheless been argued convincingly, despite this lack of explicit discussion that the stomach is in fact central to the novel. See Victor Brombert, *The Novels of Flaubert*, 49.

[7] Gustave Flaubert, *Madame Bovary*, 396. This edition hereinafter *MB*.

[8] *Madame Bovary*, Ms, Copiste (hereinafter C), f° 202; S. A. D. Tissot, *De la Santé des Gens de lettres*, 168, on reading as cause of "maladies nerveuses"; and *L'Onanisme*, 241, on the softness of beds. See also Bertrand Marquer, *L'Autre siècle de Messer Gaster?* According to Marquer (p. 100), good digestion was seen as a sign of "la santé littéraire" ["literary health"].

that Emma procures arsenic from his shop—enumerates the nervous and digestive dangers of reading:

> The optic nerve, continually obliged to convey sensations to the brain, disturbs it, causes commotion in it. Following on from this, there are palpitations, feelings of disgust, loss of appetite, poor digestion, trouble in the nerves; sleep, if it comes about, is continually agitated by epistomachiae, in other words nightmares.[9]

Homais concludes that the only way of investigating such matters would be to "examine the relations between the moral and physical and how literature and artistic labours are linked to physiology."[10] He doubtless alludes here to a significant text on digestion, Cabanis's *Rapports du physique et du moral de l'homme* (1802). He affirms moreover a connection between discursive matter or practice and the body, a connection expressed in the neologism "épistomachies" ["epistomachiae"], seemingly synonymous with nightmares.[11] One reading of this term is that the pharmacist substitutes the Greek "gastrie" for the Latin "stomachie," in an ostentatious alternative to "épigastrie," affirming a specifically digestive consequence of reading. But why should stomach complaints be synonymous with nightmares?

Early nineteenth-century medical literature on nightmares identified the stomach as a site of importance.[12] The causes of nightmares were grouped with "those likely to cause gastric affections" ["celles qui disposent aux affections gastriques"] and highlighted the susceptibility to nightmares of "ceux chez qui l'on observe quelques signes d'embarras gastrique" ["those in whom some signs of gastric discomfort can be observed"].[13] The first monograph devoted to the nightmare as exemplar of "maladies nerveuses" ["nervous maladies"] identified "a violent

[9] "Le nerf optique, continuellement obligé de porter au cerveau les sensations, l'ébranle, il le commotionne. [...] – et de là, palpitations, dégoûts, perte de l'appétit, les digestions se font mal, l'innervation se trouble; [...] le sommeil, s'il se présente, est continuellement agité par des épistomachies autrement dit cauchemars [...]." C, f° 202.

[10] "examiner les rapports du moral et du physique et comment la littérature et les travaux artistiques se rattachent à la physiologie." Ibid.

[11] Pierre-Jean-Georges Cabanis, *Rapports du physique et du moral de l'homme.*

[12] See Duffy, *Flaubert, Zola and the Incorporation of Disciplinary Knowledge*, 55–57, 127–133.

[13] H. Petit, "Incube," 306.

pain in the epigastric region" ["une violente douleur dans la région épigastrique"] as a symptom accompanying nightmares, viewed as a phenomenon that "the poor state of the stomach is likely to cause" ["le mauvais état de l'estomac est susceptible d'occasionner"].[14]

In linking nightmare and stomach, Homais dispenses contemporary medical wisdom. However, another key feature of historical and contemporary medical accounts and cultural representations of the nightmare was the struggle between the subject and a beast, or demonic creature, typically seated on the subject's chest or stomach. Homais's nightmare synonym, "épistomachie" ["epistomachia"], might in an alternative reading [as "epistomachy"] be understood as a killing ("machie"; "machy") of, or struggle with, knowledge or science ("epistêmê").[15] Knowledge would thus be the beast being grappled with by the subject, and indeed, in Emma's situation the problem identified is a debilitating struggle with knowledge or discourse, implicitly poorly digested by a subject incapable of ordering it according to contemporary taxonomic imperatives.

Another related discourse present in *Madame Bovary* is orthopaedics, which itself, while being primarily concerned with putting bodily deviations in order, has digestive aspects. Vincent Duval, author of the manual recommended by Homais and consulted by Charles for the disastrous club-foot operation on Hippolyte the stable boy, intended to put his wayward limb in order, asserted "the most obvious connections between gastro-intestinal affections, inflammations of the cerebro-spinal apparatus and limb deformities."[16] These conditions are further linked to another condition on which Duval is a specialist, scrofula, a disorder of the lymphatic system, "the principal instrument of interstitial absorption effecting the decomposition of bodies."[17] Duval's *Traité théorique et pratique de la maladie scrofuleuse* (1852) proposed dietary regimens and treatments for scrofula in general, but counselled explicitly against using them for chronic ophthalmic varieties.[18] Homais, having identified

[14] L. Dubosquet, *Dissertation sur le cauchemar*, 5, 7.

[15] See Anon., "Un jeu de mot de Flaubert?" l.

[16] "les rapports les plus évidens entre les affections gastro-intestinales, les inflammations de l'appareil cérébro-spinal et des difformités des membres." Vincent Duval, *Traité pratique du pied-bot*, 296–297.

[17] "principal instrument de l'absorption intersticielle qui effectue la décomposition des corps." Ibid., 6.

[18] Ibid., *Traité théorique et pratique de la maladie scrofuleuse*, 445.

"une affection scrofuleuse" ["a scrofulous affection"], advises the Blind Beggar to "suivre un régime" ["follow a regimen"]: "good wine, good beer, good roast meats" ["de bon vin, de bonne bière, de bons rôtis"] (*MB*, 373–374). Even if this prescription is inappropriate given his chronic and localised symptoms, it is yet another aspect of the digestive dimension of his symbolic presence, linked with cultural markers of the nightmare as they manifest themselves in the novel: equinity, deformity, sudden waking and *douleurs gastriques* [stomach pains]. As Emma sleeps in the stagecoach, for example, the nightmarish blind man grapples with the vehicle, "fully squatting like an ape on the running board" ["tout accroupi comme un singe sur le marchepied"].[19] Not only does he resemble the simian demon who presses on the subject's chest or abdomen during the nightmare; he is also in a defecatory position.

Flaubert's novel conveys nightmarish monstrosity through the inclusion of nightmare symptoms found in the cultural imaginary and in medical discourses. As in relation to other disciplinary domains, it nods to digestive aspects of orthopaedics, dermatology and nervous illnesses, linking them to the digestion, that is the appropriate ordering, of knowledge. Knowledge is, indeed, at the heart of the various personal, professional and discursive struggles that *Madame Bovary* articulates, for which the *épistomachie*, the nightmarish struggle with knowledge itself, symbolically represented by Homais's and Emma's struggles with the deformed and equine Aveugle, provides an emblematic model.

Zola

For Susan Harrow, Zola's corpus is emblematic of a "relationship between the writing of the body and the body of writing."[20] It is unsurprising, then, that Zola should be an irrepressible user of incorporative, indeed digestive, metaphor in a project that ultimately reflects upon the accumulation and ordering of knowledge. *Le Ventre de Paris* [The Belly of Paris] (1873), discussed elsewhere in this volume, is archetypical of Zola's explicit use of digestive metaphor; the novel of the market district functioning as stomach establishes digestive oppositions between *gras* [fat] and *maigres* [thin], mirroring social ones between rich and poor.

[19] *Madame Bovary*, Ms, Brouillons, vol. 5, f° 228v.

[20] Susan Harrow, *Zola: The Body Modern*, 14, 16.

Elsewhere in Zola's oeuvre, incorporation is less explicitly digestive. For Laura Otis, Zola's project is ultimately "centred around an image of accumulation." Accumulation is a subject, but also the practice of the novelist concerned with the convincingly plausible ordering of information in a cycle of documentary fictions: "Accumulation, memory and reproduction [...] characterize the narrative structure, as each novel draws upon previous ones and lays the foundation for future ones."[21] This is evident in Zolian novels which do not use explicitly digestive metaphors. *Nana* (1880), for instance, charts the rapacious material consumption of its antiheroine, pandering to the insatiable sexual appetites of "a whole society rushing after ass" ["toute une société se ruant sur le cul"] and ultimately consumed by *pourriture*.[22] *La Bête humaine* (1890) represents a railway network as being "like a great body, a giant being" ["comme un grand corps, un être géant"] incorporating the national territory and component metaphorical body parts, in a text—incorporating psychiatric discourse—which charts the incorporation within the physiological body of hereditary mental illness, previously considered within the metaphysical domain.[23]

Other novels by Zola exploit the maximum metaphorical potential from digestion to illustrate accumulative capitalist processes. *Au Bonheur des Dames* [The Ladies' Paradise] (1883) represents the expansion of the department store in terms of accumulative engulfment, and its inner workings represented in mechanically digestive terms. While the store engulfs its Parisian environment commercially and physically, effective in transforming Paris because it is "made from the flesh and blood of woman" ["faite de la chair et du sang de la femme"] (*RM* III, 460), it swallows not only its surroundings and its female *clientèle*, but also consumer goods, in "a continual act of engulfment" ["un engouffrement continu"], generating "un ronflement de rivière" ["the rumbling of a river"] (*RM* III, 462). This functions alongside other liquid imagery as part of the "continuel flot des marchandises" ["continual flood of goods"] (*RM* III, 610) through a constantly functioning digestive system, paralleled by that of a bestialised Paris, emitting "the snoring of a

[21] Laura Otis, *Organic Memory: History and the Body in the Late Nineteenth and Early Twentieth Centuries*, 58.

[22] Émile Zola, Ms, *Nana*, *dossier préparatoire*, f° 208.

[23] Zola, *La Bête humaine* [1890], in *Les Rougon-Macquart*, IV (1966), 1035. This edition hereinafter *RM* followed by volume and page number.

sated ogre, digesting the fabrics and sheets, the silks and laceworks, on which he had been force-fed since the morning" ["un ronflement d'ogre repu, digérant les toiles et les draps, les soies et les dentelles, dont on le gavait depuis le matin"] (*RM* III, 499).

L'Argent [Money] (1891) shares expansionist imagery with *Au Bonheur des Dames*, charting the spread of colonialism through infrastructural networks and concurrently showing the workings of capitalist accumulation through the expansion of the financial institution funding it: "'The Banque Universelle: it's taking in everything; it's covering the world'" ["'la Banque Universelle, [...], ça englobe tout, ça couvre le monde...'"] (*RM* V, 82). It contains allusions to digestion, such as in an explanation of how capital will eat itself, "the great producers devouring the workers, the great credit houses and the department stores killing any competition, getting fat on the ruin of the little banks and little shops."[24] This extract, digested from a Marxist primer, represents class conflict in digestive terms, echoing *Au Bonheur des Dames* in its enactment of the destructive ingestion of small institutions by larger ones, but also offering a variation on the theme of *gras* consuming *maigres* through the use of the verb "engraisser" ["to fatten"].[25] Such accumulation is linked to excremental product, with financial speculation seen as "the human excess, the essential manure, the compost from which progress sprouts" ["l'excès humain, l'engrais nécessaire, le fumier sur lequel pousse le progrès"] (*RM* V, 382). In the capitalist digestive cycle, "l'horrible argent, qui salit et dévore" ["horrible money, which dirties and devours"] (*RM* V, 216), is "le fumier dans lequel poussait l'humanité de demain" ["the manure in which the humankind of tomorrow was growing"] (*RM* V, 398). The likening of money to excrement as integral element of the financial circulus here may be contrasted with the situation of excrement *outside* the capitalist economy in *La Terre* (1887).[26]

The novel of accumulation in which Zola deploys digestive metaphor most powerfully, however, is *Germinal*, in which we also see a tentative linking of digestion with the assimilation of discourse. This novel of class conflict deploys the *gras/maigres* oppositions of *Le Ventre de Paris*: "les

[24] "les grands producteurs dévorant les ouvriers [...], les grandes maisons de crédit et les grands magasins tuant toute concurrence, s'engraissant de la ruine des petites banques et des petites boutiques [...]" (*RM* V, 43–44).

[25] Albert Schaeffle, *La Quintessence du socialisme*.

[26] See Manon Mathias, "Recycling Excrement in Flaubert and Zola."

longues digestions heureuses" ["the long, happy digestions"] of the rich contrast with the difficult digestion of the poor: "meat was good; but they digested it poorly as they saw it too rarely" ["[c]'était bon, la viande; mais ils la digéraient mal, ils en voyaient trop rarement"].[27] Like *Au Bonheur des Dames*, *Germinal* features an "ogre repu," either the mine insatiably swallowing and grinding human flesh, or the abstract and occulted sated ogre of capital, engaged in its "painful digestion of human flesh" (60) ["digestion pénible de chair humaine"]. Satisfaction of the permanently ravenous beast's appetites is regulated, as if according to a diet: at the day's end, "the devouring pit had swallowed its daily ration of men" ["[l]e puits dévorateur avait avalé sa ration quotidienne d'hommes"] (84).

The scene of most of these digestive processes is the "entrailles du sol" ["entrails of the soil"] (343). The mine explodes on account of its "entrailles trop grêles" ["entrails that were too spindly"] (557); its volatility lies in its entrails; so too do human volatility and violence. Etienne's surging hereditary killer instinct is presented thus:

> An abominable voice, within him, deafened him. It rose up from his entrails, beat inside his head with hammer blows, an urgent craze for murder, a need to taste blood.[28]

Stirring in Etienne's bowels is an urge not just to kill, but to ingest. Elsewhere, Etienne's vision of a Darwinian struggle is to "see men devour one another until the end of time" ["voir les hommes se manger entre eux jusqu'à la fin du monde"] (292). In one crypto-Freudian reading, the synonymity of verbs of eating ("manger" ["eat"], "dévorer" ["devour"]) with killing reflects an "equivalence [...] between digestion and death." Moreover, the sustained repetition of "accroupi" ["squatting"] to characterise the sated ogre of capital, "gluttonous beast, squatting there in order to eat everyone" ["bête goulue, accroupie là pour manger le monde"] (52), symbolises "the nauseous posture of an anthropomorphic monster in the process of defecating" as well as about to eat/kill its prey in a process of "aggression-digestion-*defecation*."[29]

[27] Émile Zola, *Germinal*, 122, 199. Hereinafter referred to by page number only.

[28] "Une voix abominable, en lui, l'assourdissait. Cela montait de ses entrailles, battait dans sa tête à coups de marteau, une brusque folie du meurtre, un besoin de goûter au sang" (465).

[29] David Bellos, "From the Bowels of the Earth."

The mine itself merely swallows workers; it is the "impersonal god, unknown to the workers" ["dieu impersonnel, inconnu des ouvriers"] occulted by the infrastructure of the mine, which enacts the extractive processes of digestion and excretion, "squatting somewhere, in the mystery of its holy of holies, from where it sucked the life out of the starving hordes who nourished it!" ["accroupi quelque part, dans le mystère de son tabernacle, d'où il suçait la vie des meurt-de-faim qui le nourrissaient!"] (344).

The bourgeoisie as well as hungry workers fear a devouring beast; Deneulin accuses his fellow capitalists of increasing the likelihood of popular unrest by making concessions: "'You're sharpening the monster's teeth so it can devour us.'" ["'Vous aiguisez les dents du monstre pour qu'il nous dévore.'"] (262). Etienne's intellectual needs in the class struggle are rendered in similarly digestive terms; he is "devoured above all by a need for knowledge" ["dévoré surtout du besoin de savoir"] (195). He acquires books by mail order (including "*l'Hygiène du mineur*"); their "badly digested reading left him exalted" ["lecture mal digérée acheva de l'exalter"] (215). This intrusion of "digestion" into the realm of reading and the text is no mere turn of phrase; poor digestion in a literal sense elsewhere causes similar effects:

> When the Maheus and the others had digested their soup of clear water too quickly, they were lifted into a semi-vertiginous state, the ecstasy of a better life.[30]

Over-rapid digestion has the same effect in both cases. A later example of the dizzyingly ecstatic effects of the unordered assimilation of sketchy knowledge is couched in firmly dietary terms:

> His badly digested reading came back to him: examples of peoples who had burned down their cities to stop the enemy, vague histories of men allowing themselves to die of starvation rather than to eat the bread of tyrants. This put him in a state of exaltation.[31]

[30] "Lorsque les Maheu et les autres avaient digéré trop vite leur soupe d'eau claire, ils montaient ainsi dans un demi-vertige, l'extase d'une vie meilleure […]" (278).

[31] "Des lectures mal digérées lui revenaient, des exemples de peuples qui avaient incendié leurs villes pour arrêter l'ennemi, des histoires vagues […] où les hommes se laissaient mourir d'inanition, plutôt que de manger le pain des tyrans. Cela l'exaltait […]" (286–287).

This exaltation is matched by confusion and lack of full understanding of what Etienne voraciously consumes (but slowly digests without arrangement into any coherent and comprehensible order):

> Although he continued to educate himself, devouring everything, his lack of method made assimilation very slow; such confusion came about that he ended up knowing things that he had not understood.[32]

Digestion in its literal and metaphorical senses is linked to that of discourse, which has visceral and spiritual effects: at a secret forest assembly, the members of Etienne's audience "no longer felt the cold; these ardent words had warmed them to their entrails. A religious exaltation raised them from the earth" ["ne sentaient plus le froid, ces ardentes paroles les avaient chauffés aux entrailles. Une exaltation religieuse les soulevait de terre"] (340).

Having sufficiently redigested the already digested versions of (presumably) Marx, Etienne turns to Darwin.

> He had read fragments of it, summarised and vulgarised in a five-*sou* volume, and, from this poorly understood reading, he concocted a revolutionary idea of the struggle for life, the thin eating the fat, a strong people devouring a feeble bourgeoisie.[33]

It is striking that Etienne comes to this digestive conclusion through poor digestion, incoherent ordering, of information—both his own and that of the edition he has read. The editorial and publishing process is a digestive one, producing "fragments" circulated on the mass market. The production and dissemination of reading material are only possible under the same industrial modernity which makes le Voreux possible. *Germinal* is in this sense a pivotal novel. In it can be seen the beginnings of a commentary on the accumulation and satisfactorily plausible arrangement of discourse.

[32] "S'il continuait à s'instruire, dévorant tout, le manque de méthode rendait l'assimilation très lente, une telle confusion se produisait, qu'il finissait par savoir des choses qu'il n'avait pas comprises" (279).

[33] "Il en avait lu des fragments, résumés et vulgarisés dans un volume à cinq sous; et, de cette lecture mal comprise, il se faisait une idée révolutionnaire du combat pour l'existence, les maigres mangeant les gras, le peuple fort dévorant la blême bourgeoisie" (340).

This commentary is given full expression in *Le Docteur Pascal*, a conclusion to the *Rougon-Macquart* cycle offering self-reflexive consideration of how the literary text incorporates extraliterary material, and of the effects of such epistemological stimuli and their ordering upon its organically conceived coherence. Pascal Rougon, the novel's eponymous hero, is engaged in the process of accumulating family knowledge, building up a *Rougon-Macquart* family archive consisting of dossiers documenting the hereditary traits of its individual members, represented visually on a family tree: his purpose in doing so is to seek out the specific origin of his own inherited characteristics. He is simultaneously engaged in an experimentative search for tonic cures, initiated when he comes across a fifteenth-century medical text prescribing the use of an extract from animal organs:

> The doctor, reading an old medical book from the fifteenth century, was very struck by a medicinal treatment known as "medicine of signatures." To heal a sick organ, it sufficed to take from a sheep or from a cow the same healthy organ, to boil it, then to have the patient swallow the stock. The theory was to repair using what was similar, and for illnesses of the liver especially, said the old book, the cures were countless.[34]

This treatment, the name of which expresses an explicitly scriptive dimension, is specific and targeted. Pascal later abandons it for something quite unspecific: a new therapeutics based on revitalisation of the body through hypodermic injections premised on the non-specificity of the substance injected. This idea, drawn from a work on hypodermic injections and transfusions by a Dr Jules Chéron, is at the novel's conceptual heart and functions as a metaphor for what the novelist has been doing throughout the development of the *Rougon-Macquart* cycle: injecting contemporary discourse into the fictional text, incorporating it, blending it creatively to produce a coherent textual organism of interlinked components.[35] As with Pascal's hypodermic injections, the nature

[34] "[L]e docteur, lisant un vieux livre de médecine du quinzième siècle, fut très frappé par une médication, dite 'médecine des signatures.' Pour guérir un organe malade, il suffisait de prendre à un mouton ou à un bœuf le même organe sain, de le faire bouillir, puis d'en faire avaler le bouillon. La théorie était de réparer par le semblable, et dans les maladies de foie surtout, disait le vieil ouvrage, les guérisons ne se comptait plus" (*RM* V, 948).

[35] Duffy, *Flaubert, Zola and the Incorporation of Disciplinary Knowledge*, Ch. 6; Jules Chéron, *Introduction à l'étude des lois générales de l'hypodermie*.

of the discursive substance ingested into the textual body is immaterial: its curative effect on the "jeu de l'organisme" ["the interplay within the organism"], on bodily integrity and coherence (or "tonicité" ["tonicity"], as both Pascal and Chéron put it), both internally among its component parts, and with the exterior environment, is what counts.[36] The culminating *Rougon-Macquart* novel thus rejects specificity as means of healing or understanding the physiological, familial or textual body. It represents a meditation on the relationship between the literary text—implicitly analagous with the body—and what it ingests from its ambient discursive milieu.

Huysmans

Huysmans has a well-documented preoccupation with digestion.[37] The two works to be discussed here explore similar textual preoccupations to those explored in *Le Docteur Pascal.* Des Esseintes in *A Rebours* longs for a form of literature that can be extracted, condensed; Durtal in *En Route* initially seeks readily digestible spiritual nourishment, before a dietary and spiritual regime convinces him that it is the integrity of the receiving digestive system that counts.

À Rebours [Against Nature] (1884), as is argued elsewhere in this volume, establishes a link between writing and digestion through engagement with contemporary health discourse on digestion and its commercial manifestations. The digestive aspect of Des Esseintes's concerns with literature is no doubt related to his own bodily concerns with digestive discomfort (in the form of "dyspepsie nerveuse" ["nervous dyspepsia"]).[38] He has a generic preference for the prose poem, as it offers a condensed, digestible form of literature—"à l'état d'of meat" ["in the form of extract of meat"]—that does not require the discomfort of reading long, overdone, disorderly descriptions. The problem becomes one of writing.

[36] "la tonicité, c'est à dire [la] vitalité" ["Tonicity, That Is, Vitality"]. Chéron, 538; "la tonicité": "l'état de santé parfaite" ["Tonicity: The State of Perfect Health"]. *RM* V, 1177–1178.

[37] See Francesco Manzini, "Nutrition, Hunger and Fasting"; Jean Borie, *Huysmans*, 145–190.

[38] Joris-Karl Huysmans, *À Rebours*, 263. Hereinafter *AR*.

> A novel concentrated in a few sentences which would contain the condensed meat of the hundreds of pages still employed to establish milieu, to delineate characters, to pile up observations and trivial facts.[39]

What the proposed novelistic problematics entail is in fact a rejection—in favour of a distilled or extracted "suc" or "osmazome"—of "entassement" ["piling up"] or messy accumulation of the raw material of fiction, and the separation of accumulation from digestion, taking place prior to consumption so that it is the producer, the writer or the editor, who digests and orders, rather than the eventual consumer, the reader, just as the consumer of a condensed meat distillation avoids the discomfort of digestion since the manufacturer has extracted the essential nutritional content.[40] *À Rebours* advocates a new kind of writing practice, privileging neat, pre-ordered concision over disorderly accumulation within the textual product. In its rejection of the accumulation characteristic of the realist or naturalist novel, it also apparently rejects what it depicts, namely the implicitly messy accumulation of objects by its protagonist.

Literature itself is conceived of in organic terms, as an ailing body suffering from the rich substances it contains and from its inability to "exprimer" ["express"], that is to excrete, everything that it has been forced to assimilate:

> The decadence of a literature, irreparably attacked within its organism, weakened by the age of ideas, exhausted by excesses of syntax, sensitive only to the curiosities which make the sick feverish and yet hastening to express everything in its decline, bent on wishing to rectify all omissions of joy, on bequeathing the most subtle memories of pain, on its death bed, had become incarnate in Mallarmé.[41]

[39] "un roman concentré en quelques phrases qui contiendraient le suc cohobé des centaines de pages toujours employées à établir le milieu, à dessiner les caractères, à entasser à l'appui les observations et les menus faits" (*AR*, 260).

[40] Homais, who (*MB*, 162) vaunts his knowledge of "osmazome, suc et gélatine" ["Osmazome, Concentrate and Gelatine"], is also an expert at disseminating condensed, re-ordered, discourse for the consumption of the local populace.

[41] "[L]a décadence d'une littérature, irréparablement atteinte dans son organisme, affaiblie par l'âge des idées, épuisée par les excès de la syntaxe, sensible seulement aux curiosités qui enfièvrent les malades et cependant pressée de tout exprimer à son déclin, acharnée à vouloir réparer toutes les omissions de jouissance, à léguer les plus subtils souvenirs de douleur, à son lit de mort, s'était incarnée en Mallarmé" (*AR*, 261).

Mallarmé is privileged as an alchemist capable of expressing through the prose poem the literary body's decline. The tentative literary method proposed, moreover, is not just one involving distillation and extraction. There is also, through the involvement of "un alchimiste de génie" ["an alchemist of genius"], "un magique écrivain" ["a magical writer"] and an appropriately receptive "idéal lecteur" ["ideal reader"] (*AR*, 260) intellectually capable of digesting the sublime extract, an occult or spiritual dimension.

This idea is developed in *En Route* (1894), second instalment in the Durtal tetralogy and the first of three charting its protagonist's conversion, which from its outset establishes an analogy between the soul and the body, or more specifically, between the soul and the digestive system. This is hardly surprising, as the novel's central drama is that of the repentant philanderer Durtal entering a monastery and undergoing a regime that is at once digestive and spiritual. The digestive regimen involves fasting and ascetic abstinence from luxurious foods, including meat; the spiritual one has a strong digestive dimension, in that the subject receives the word of God through a spiritual digestive process. Early in the novel, Durtal mulls over the "moyen" ["means"] of his conversion, "the one that the saviour has used for me. And this one consists in I know not what; it is something analagous to the digestion of a stomach working without one feeling it." ["celui dont le sauveur s'est servi pour moi. Et celui-là consiste en je ne sais quoi; c'est quelque chose d'analogue à la digestion d'un estomac qui travaille, sans qu'on le sente."] (*ER*, 325).[42] Durtal at this stage is however of the view that spiritual digestion—like physical digestion—is no easy matter, especially if the word of God is expressed in terms of indigestibility. At St-Sulpice, in the novel's opening scene (*ER*, 309), a Parisian priest is characterised negatively in terms of the viscosity of his enunciation ("from the vaseline of his delivery, from the grease of his accent" ["à la vaseline de son débit, à la graisse de son accent"]) and his "solidly nourished" ("solidement nourri") state to which it is implicitly linked. His utterances are perceived as indigestible because of their texture, and foreshadow Durtal's dietary regime at the monastery, where "he absorbed painfully a purée of potatoes in hot oil resembing vaseline" ["il absorbait péniblement une purée de pommes de terre à l'huile chaude qui ressemblait [...] à

[42] Huysmans, "En Route," in *Le Roman de Durtal*, 325. Hereinafter *ER*.

de la vaseline"] (*ER*, 528). Urban clerics like this priest are likened to mediocre restaurateurs, and what they serve up is compared unfavourably to desirable spiritual nutrition:

> If these grubby caterers to the soul had any talent, if they served their lodgers fine foods, essences of theology, coulis of prayers, solidified concentrate of ideas, they would be left to vegetate without being understood by their flock.[43]

What is desirable, in fact, and privileged by Durtal, at this stage, is sublime spiritual food that does not require significant digestive effort, like the literary *osmazome* privileged by Des Esseintes, because it has been extracted in as pure a form as possible from its source material. Paradoxically, the desired ease of digestion is, as for Des Esseintes's "ideal reader," only available to an elect, as opposed to the uncomprehending masses.

Even liturgy Durtal admires, such as the *De Profundis* recounting the "boulimies charnelles" ["carnal bulimias"] of King David (*ER*, 312), causes digestive discomfort:

> This sublime request, ending in sobbing at the moment when the soul of the voices was on the point of transcending the boundaries of the human, twisted Durtal's nerves, made his heart shudder.[44]

Durtal's digestive turmoil at such liturgy is unsurprising given that his horror of life in general is expressed in alimentary terms as "this disgust with existence" ["ce dégoût de l'existence"].

Elsewhere, liturgy has the capacity to be incorporated within the physiological body, as in the case of Aquinas's plain-chant "which had penetrated him to the marrow, which was joined with him body and soul" ["qui l'avait pénétré jusqu'aux moelles, [...], qui faisait corps et âme avec lui"] (*ER*, 384). Liturgical discourse can also take on the metaphorical

[43] "Si ces gargotiers d'âmes avaient du talent, s'ils servaient à leurs pensionnaires des nourritures fines, des essences de théologie, des coulis de prières, des sucs concrets d'idées, ils végéteraient incompris des ouailles" (*ER*, 309–310).

[44] "Cette requête sublime finissant dans les sanglots au moment où l'âme des voix allait franchir les frontières humaines tordit les nerfs de Durtal, lui tressailla le cœur." *ER*, 311. Note that in French "cœur," while literally denoting the heart, can also refer to the stomach.

properties of a physical substance, more specifically of a purgative reagent that interacts chemically with what it comes into contact with. Its effect on the Madeleine during funerals is described thus:

> Like a superterrestrial antiseptic, like an extrahuman thymol, liturgy purges, disinfects the impious ugliness of these premises.[45]

If liturgy has chemical properties, hagiographical discourse too can be extracted into effective essences presumably digestible by the receptive soul of the convert, such as "the mournful concentrate and the white perfume of the very ancient translations of Varagine's *Golden Legend*" ["le suc dolent et le blanc parfum des très anciennes traductions de la *Légende dorée* de Voragine"] metaphorised as "those plaintive flowers cultivated by monks in the precincts of monasteries" ["ces fleurs plaintives que les moines cultivèrent dans les pourpris des cloîtres"] (*ER*, 329), elements of a discursive pharmacopoeia.

Liturgy clearly has curative properties. The *Stabat* in St-Sulpice, "that music of aching moans" ["cette musique de plaintes endolories"], consoles the Virgin Mary "by pressing upon the wound some of that prose yielding water and blood like the wound of Christ" ["en pressant sur la blessure de cette prose qui rendait de l'eau et du sang comme la plaie du Christ"] (*ER*, 333). The Church, source of such liturgy, is also a source of sustenance and therapy for the soul, as Durtal exclaims:

> [The Church] is the divine stud farm and the celestial clinic for souls; it is she who feeds them, who rears them, who tends to their wounds [...]. Yes, but in that case, then, her prescriptions should be followed and the sacraments she demands practised![46]

What can be seen here are the beginnings of a realisation on Durtal's part that the soul as digestive system has to put itself in order through a form of spiritual dietary regime. However, Catholicism—especially if "tepid and floating, continually reheated in a *bain-marie* of false zeal"

[45] "Comme un antiseptique supraterrestre, comme un thymol extrahumain, la liturgie épure, désinfecte la laideur impie de ces lieux" (*ER*, 319).

[46] "[L'Église] est le haras divin et le dispensaire céleste des âmes; c'est elle qui les allaite, qui les élève, qui les panse [...]. Oui, mais alors, il faudrait suivre ses prescriptions et pratiquer les sacrements qu'elle exige!" (*ER*, 342)

["tiède et flottant, continuellement réchauffé par le bain-marie d'un faux zèle"] (*ER*, 343)—is blandly unappealing on its own. Durtal, as a result of a shift in the subject of his sexual appetites, suffers from "a kind of dyspepsia of the soul, no longer digesting the customary subjects, searching, in order to feed himself, for *ravigotes* of daydreams, seasonings of ideas; it appeared thus that it was this lack of appetite for healthy meals which had brought about this desire for baroque dishes, this murky ideal, this desire to escape from the ego, to transgress, even if only for a second, the accepted limits of the senses."[47] Digestive analogies applicable in the sexual sphere apply in the spiritual sphere also so that Catholicism, as simultaneous "révulsif" ["revulsant"] and "déprimant" ["depressant"], would only make Durtal, whose liturgical tastes are epicurean rather than bland, a hostage to "the agitation of his nerves" ["l'émoi de [s]es nerfs"]. Mere priests, moreover, "will propose imbecilic readings for me; they'll pour me out bowlfuls of their pious veal *bouillon*" ["me prôneront des lectures imbéciles; ils me verseront à pleins bols leur bouillon de veau pieux!"] (*ER*, 348).

This conclusion prompts Durtal to consult the Abbé Gévresin, a fellow enthusiast for the ascetic martyr Sainte Lydwine. His distinguished spiritual advisor states explicitly that he is analagous with the physician. What he prescribes is similar to Pascal Rougon's "dynamic therapeutics," in that the desired cure is to be found not in a specific substance or sacrament—"I am not prescribing you any religious remedy" ["Je ne vous prescris aucun remède religieux"] (*ER*, 375)—but within the soul, analagous to the body, of the patient, prescribed instead "some precepts of pious hygiene" ["quelques préceptes d'hygiène pieuse"] (*ER*, 376), analagous to the maintenance of the body through good digestive practice, "hygiène alimentaire" ["alimentary hygiene"]. Thus begins a gradual shifting of focus from the extractive to the holistic, and from the substance consumed to the system receiving it. Durtal begins coming round to the realisation that what requires putting in order is the soul, rather than liturgy.

[47] "une sorte de dyspepsie d'âme, ne digérant plus les sujets coutumiers, cherchant pour se nourrir des ravigotes de songeries, des salaisons d'idées; ce serait donc cette inappétence des repas sains qui aurait engendré cette convoitise de mets baroques, cet idéal trouble, cette envie de s'échapper hors de moi, de franchir, ne fût-ce que pendant une seconde, les lisières tolérées des sens" (*ER*, 346).

Initially, Durtal, "discontented like those patients who take against doctors when they only prescribe pale drugs" ["mécontent de même que ces malades qui en veulent aux médecins lorsque [...] ceux-ci ne leur ordonnent que de pâles drogues"], is sceptical of the abbé's prescriptions: "this fine fellow of a priest is a practitioner of expectant medicine; instead of stopping my attacks through energetic remedies, he recommends that I go to bed early, to avoid catching cold."[48] It is clear that what is being proposed is a regime, a spiritual diet. One such regime is a "système" ["system"] for keeping potentially wayward sinners in line: "purge them with the Sacred Species in high doses" ["les épurer avec les Saintes Espèces prises à de hautes doses"] (*ER*, 377); the sacred purgative matter to be consumed is medicalised. But rather than encourage Durtal simply to receive the sacraments as a means of resisting temptation, the priest proposes a diet of abstinence so that he will find within himself "that hunger for the Eucharist" ["cette faim de l'Eucharistie"]. The priest boils his advice down to: "cook in your juice and wait" ["cuisez dans votre jus et attendez"]. While there is clearly a culinary metaphor here, there is also an instruction to Durtal to accept his ambient milieu. Rather than ingest specific spiritual cures such as the Sacred Species, it is more important to ensure a healthy spiritual digestive system so that he might properly "absorb faith" ["absorber la foi"] (*ER*, 377). The life that Durtal is about to enter is not one of consumption, but "la vie purgative" ["the life purgative"] (*ER*, 384).

It is not, therefore, the content of diet that matters, but a disciplined regime revitalising the soul in order to make it more easily digest the word of God. Thus, the role of any monastic administrator—such as the Benedictine abbess of Gevresin's acquaintance praised as a "regulator of souls" ["régisseuse d'âmes"]—is not to nourish souls, but to regulate them so that they will be able to digest eventual nourishment.

Durtal's impending retreat is analagous to a visit to a sanatorium. His purpose in entering a Trappist monastery is, as the abbé says, "to cure yourself there" ["pour vous y curer"]. To this end, Gévresin entreats him: "Parlons hygiène." ["Let us talk hygiene."] (*ER*, 432). He provides detailed information about the daily routine and the regimented spiritual and physical diet to be endured (*ER*, 433). The food is bland, yet, as

[48] "ce brave homme de prêtre pratique la médecine expectante; au lieu de me couper par des médicaments énergiques mes crises, il [...] me recommande de me coucher de bonne heure, de ne pas attraper froid..." (*ER*, 376).

Durtal will find, digestible. The regimen, which is not only dietary but includes immersion in the ambient environment, is a tonic for numerous complaints, including digestive ones: "'[Y]ou complain of stomach pains, migraines, ills of the entrails! Well, this regime, in the countryside, in the fresh air, will cure you better than the drugs that you are being made to take.'"[49] For the body, the treatment is a tonic one, avoiding specific, targeted drugs.

Durtal's reservations are primarily dietary ones, focused on specific substances:

> But it would be all the more necessary to know whether the body is in a state to withstand such a remedy; [mine] weakens when it is not comforted by the blood of meats, and neuralgias arise, as soon as mealtimes change. I will never be able to hold up out there with vegetables cooked in hot oil or in milk; first of all, I hate oily food and I detest milk all the more for the fact that I digest it poorly.[50]

However, more concerned by the prospect of confession—"the painful extraction of shameful deeds" ["la douleureuse extraction des hontes"]—he becomes reconciled to his impending diet:

> As for the food, little matter that it should be inhuman if my stomach digests it; eating badly, getting up in the night: all that is nothing as long as the body can endure it.[51]

Durtal becomes indifferent to gastronomic specificity; important are digestion and the integrity of the digestive system. This is the conclusion he simultaneously comes to concerning his spiritual well-being.

[49] "vous vous plaignez de gastralgies, de migraines, de maux d'entrailles! eh bien, ce régime-là, à la campagne, au plein air, vous guérira mieux que les drogues qu'on vous fait prendre" (*ER*, 433).

[50] "Mais encore faudrait-il savoir si le corps est en état de supporter un remède pareil; [le mien] tombe en faiblesse quand il n'est pas réconforté par le sang des viandes et des névralgies surviennent, aussitôt que les heures des repas changent. Jamais je n'arriverai à tenir là-bas avec des légumes cuits dans de l'huile chaude ou dans du lait; d'abord, je déteste la cuisine à l'huile et j'exècre d'autant plus le lait que je le digère mal" (*ER*, 437).

[51] "Quant à la cuisine, peu importe qu'elle soit inhumaine si mon estomac la digère; mal manger, se lever dans la nuit, ce n'est rien, pourvu que le corps l'endure" (*ER*, 440).

This realisation is due partly to repeated emphasis on parallels between soul and body, spiritual and physical nourishment, faith and medicine. Gévresin reminds him, for example, that the Eucharist is "a spiritual medicine; we go to the Saviour as we go to a doctor; we bring our soul for him to treat and he treats it!"[52] The spiritual regime is an unpleasant purgative medicinal extract to be consumed in full: "the colocynth has to be drunk down to the last drop; it's all or nothing" ["le chicotin doit être bu jusqu'à la dernière goutte, c'est tout ou rien"] (*ER*, 442). Durtal finally commits himself to the retreat on attending a mass and realising its tonic value in preparing him for an eventual act of consumption of the Eucharist: "This mass strengthened him as would a tonic" ["Cette messe le renforça comme un tonique"] (*ER*, 452). Durtal acknowledges explicitly that the cure sought for his soul is not a targeted one, but one akin to that sought in a spa, focused on the whole body: "there are plenty of people who go to Barèges or Vichy to take cures of the body; why shouldn't I go to take a cure of the soul in a Trappist monastery?"[53]

Following this decision, Durtal endures further dietary anxieties and physical symptoms: following an anxious evening, "[he] woke up ill; a furious neuralgic pain was boring into his temples; he tried to reduce it using antipyrine, but this medicine, taken in high dosage, turned his stomach without softening the hammer blows which pounded against his skull."[54] Targeted medication has negative, digestive effects. Following this experience, he anticipates nervous symptoms as a result of the diet to be undertaken at the Trappist monastery of Notre-Dame de l'Âtre: "as soon as I am badly nourished, I have neuralgia" ["dès que je suis mal nourri, je névralgise"] (*ER*, 465). In his preparations for departure, he engages in an accumulative, rather than truly digestive, practice to counter his anticipated digestive woes: "he crammed into the corners [of his suitcase] some sugar, some packets of chocolate, in order to try to

[52] "un médicament spirituel; on va au Sauveur de même qu'on se rend chez un médecin; on lui apporte son âme à soigner et il la soigne!" (*ER*, 441).

[53] "il y a bien des gens qui vont à Barèges ou à Vichy faire des cures de corps, pourquoi n'irais-je pas, moi, faire une cure d'âme dans une Trappe?" (*ER*, 453).

[54] "[il] se réveilla malade; une névralgie furieuse lui vrillait les tempes; il tenta de la réduire avec de l'antipyrine, mais ce médicament, pris à haute dose, lui détraqua l'estomac sans amortir les coups de vilebrequin qui lui térébraient le crâne" (*ER*, 463–464).

confound, if need be, the anxieties of an unfed stomach; [...] packets of antipyrine, a vial of laudanum."[55]. On arrival, having been briefed (of the monks) by his driver that "they say that their stomachs are eaten by ants" ["on dit que les fourmis leur mangent le ventre"], Durtal sees the specifically digestive aspect of his stay as being what is most likely to shorten it: "it will depend on how my body copes with the struggle; I have quite a sick stomach" ["cela dépendra de la façon dont mon corps supportera la lutte; j'ai l'estomac assez malade"] (*ER*, 483–484).

Durtal however easily adapts to the dietary regime and overcomes his digestive troubles and fear of alimentary blandness. He reads of the lives of saintly mystics who transcend digestive difficulties to receive God's grace. Sainte Angèle, for example, who had led a life of adulterous sin, finds God through overcoming increasing degrees of digestive discomfort:

> One day, an attack of retching comes to her before a leper whose raised scabs are infectious; to punish herself for her disgust, she drinks the water in which she has washed the scabs; nauseous attacks overcome her; she chastises herself further by forcing herself to swallow in one gulp a scale which the water has been unable to carry away and which has stayed in her gullet.[56]

Her reward is that Jesus "dispenses her of the need to eat, and feeds her only with the Sacred Species" ["la dispense du besoin de manger, ne la nourrit qu'avec les Espèces saintes"] (*ER*, 510). Sainte Lydwine's urge to take on the sufferings of others is characterised digestively as "la gloutonnerie des plaies" ["the gluttony of sores"] (ER, 350). In the case of the medieval abbot Pierre Monoculus, subjected to "névralgies" ["neuralgic attacks"] by the devil's temptations and cured through ascetic self-discipline, digestive matter functions as restorative cure in the manner of Pascal Rougon's *médecine des signatures*: "his entrails were washed in wine which was distributed, for this mixture constituted a powerful

[55] "[il] tassait dans les coins [de sa valise] du sucre, des paquets de chocolat, pour essayer de tromper, s'il était besoin, les angoisses de l'estomac à jeun; [...] des paquets d'antipyrine, une fiole de laudanum" (*ER*, 467).

[56] "Un jour, un haut-le-cœur lui vient devant un lépreux dont les croûtes soulevées infectent; pour se punir de son dégoût, elle boit l'eau dans laquelle elle a lavé ces croûtes; des nausées la reprennent; elle se châtie encore en se forçant à avaler une écaille que cette eau n'a pu entraîner et qui lui est restée dans le gosier, à sec" (*ER*, 510).

remedy" ["on lava ses entrailles dans du vin qui fut distribué, car cette mixture constituait un puissant remède"] (*ER*, 546).

Durtal's experience in the monastery is a positive one both physically and spiritually, precisely on account of an acceptance that the well-being of body and soul are interconnected, especially from a digestive point of view. When Durtal finally confesses, the oblate Monsieur Bruno expresses empathy in digestive terms with Durtal's presumed relief: "Eh? It's a brave load less on the stomach!" ["Hein? c'est un fameux poids de moins sur l'estomac!"] Confession is, moreover, "le grand lavage" ["the great enema"] (*ER*, 522). The soul, the spiritual stomach, is evacuated of indigestible sin. At the same time, "his whole being had in some sense been injected with divine effluvia" ["tout son être avait été, en quelque sorte, injecté d'effluves divins"] (*ER*, 540).

The aim of the purgative experience of confession is to be able to receive communion. Durtal has observed the monks returning from the altar in the chapel (the other focus of monastic life being the refectory) and has noticed that "it seemed that, pushed back by the power of the Sacrament against the walls of the body, the soul filtered through the pores and lit up the epidermis with that special light of joy."[57] The soul thus acts as a digestive filter for what it ingests spiritually before transmitting it to the rest of the body. Before Durtal can engage in this practice, however, he must do penance. Debating with himself over whether his confessor has prescribed one or ten decades of the rosary, Durtal concludes the latter, on the grounds that it is the whole exercise that counts rather than anything specific: "I found repugnant those droplets of devotion condensed into globules, so it was natural that he made me ingurgitate the rosary, at high dosage!" ["j'avais une répugnance pour ces gouttes de dévotion mises en globules, il est donc naturel qu'il m'ingurgite le rosaire, à haute dose!"] (*ER*, 530). The penance is represented in digestive, and therapeutic, terms. This appears to be a rejection of the extraction that Des Esseintes embraces in the "succulence" of literature "reduced to one drop" ["réduite en une goutte"] by the likes of Mallarmé. However, Durtal's attempt to recite ten decades results in confusion: "he was stammering, forgetting the bowls of *Paters*, and going astray in the granules of *Aves*" ["il bafouillait, oubliait les bols du

[57] "il semblait que, refoulée par la puissance du Sacrement contre les parois du corps, l'âme filtrait, au travers des pores, éclairait l'épiderme de cette lumière spèciale de la joie" (*ER*, 514).

Pater, s'égarait dans les granules des *Ave*"]. Monsieur Bruno advises that Durtal's confusion over the extent of the prescribed penance is a strategy by the devil to "disgust you with the thing that has to be practised" ["vous dégoûter de la chose qu'on doit pratiquer"]. Durtal is then able to overcome the diabolically inspired disgust and to receive the host. His act of communion causes "une torpeur absolue" ["an absolute torpor"]; he is vexed by "the far too human side of the deglutition of a God" ["le côté trop humain de la déglutition d'un Dieu"], in that "he had had the host sticking to his palate and had had to search for it and roll it like a pancake in order to swallow it" ["il avait eu l'hostie, collée au palais, et il avait dû la chercher et la rouler, ainsi qu'une crêpe, avec la langue, pour l'avaler"] (*ER*, 539). Despite these digestive difficulties, his communion is a positive experience as it is underpinned by the spiritual enema of confession and absolution. Attempting to "explain to himself the treatment that the Saviour was making him follow" ["s'expliquer le traitement que le Sauveur lui faisait suivre"], he realises that the sequence of the two sacraments had been "à rebours" ["against nature"], that "Christ had made himself perceptible to the soul, before and not after" ["le Christ s'était rendu sensible à l'âme, avant et non après"] (*ER*, 540). Again, it is the preparation of the (spiritual) digestive mechanism that counts rather than the act of consumption itself.

On inquiring as to how Durtal feels after his communion, the *père hôtelier* of the monastery does not neglect to add a digestive inquiry: "And your stomach?" ["Et votre estomac?"]. To which Durtal replies: "Why, it's doing fine, Father; I'm stupefied; I have never digested so well as I have here" ["Mais il va bien, mon père; je suis stupéfié; je n'ai jamais si bien digéré qu'ici"] (*ER*, 541). The positive digestive consequences of Durtal's regime accompany the spiritual ones analogical to them. And if a voice in Durtal's head rehearses "arguments matérialistes" ["materialist arguments"] for his improved state, including that "character changes according to whether the stomach is functioning well or badly" ["le caractère change suivant que l'estomac fonctionne bien ou mal"], he is not worried: "none of them stood up: all confused the function with the organ" ["aucun ne tenait debout: tous confondaient la fonction et l'organe"] (*ER*, 573). The coherent functioning of an organ, and by extension of the body, is thus more important than its material substance. This is plausibly a version of "the interplay of all the organs" ["le jeu de tous [les] organes"] (*RM* V, 1159) privileged by Pascal Rougon. It is Durtal's stomach that has done the work, rather than having been cured by any specific substance ingested: "My stomach has reinvigorated itself and has shown itself capable of putting up with an effort of which I would never have believed it capable, and my soul has been incapable of achieving anything, vacillating and dry, so fragile, so weak!"[58] What this assertion appears to recognise is that the stomach, and by extension the body, is not purely physical and corporal, but is capable of the spiritual effort required for digestion of both alimentary nutrition and of the "Verbe" ["Word"] (*ER*, 623). There is no clear divide or opposition between body and soul. As Durtal prepares to leave, the abbot asks him: "how have you managed with our regime?" ["comment avez-vous supporté notre régime?"] To this question, which could refer either to the dietary or spiritual regimen, Durtal replies: "I have been very lucky; I landed up here, my stomach out of order, my body sick, and the laconic Trappist meals have cured me!"[59] Although Durtal's response is an emphatically bodily one, it retains the key notion, implicit in the mention of laconic meals, which could be either alimentary or spiritual, that it is not what is consumed that cures the body or soul, but the fact of not consuming, of disciplining the stomach, the soul and their digestive functions.

If *À Rebours*, then, privileges extraction so as to avoid digestive difficulties, *En Route* proposes a regulation, a disciplining, a putting in order, of the digestive system so that the extraction of sublime and easily digested essences is not necessary. These models are operative in both alimentary and discursive spheres, if we can consider the "Verbe," the word of God in the form of liturgy, disciplinary practices associated with it and the Eucharist as word made flesh, to be discursive. Above all, body and soul are not separate.

The three novelists considered here all establish clear links between digestion and the assimilation of various forms of discourse. Flaubert, as an author whose work is emblematic of extraction, distillation and subtle articulation of contemporary discourse, may well integrate medical ideas on digestion into his corpus in order to lampoon them and any valorisation of the aggregation of knowledge, but his work is no less insistent that digestion is an important and difficult matter, whether of

[58] "Mon estomac s'est ravigouré et s'est montré apte à supporter un effort dont jamais je ne l'eusse cru capable et mon âme a été au-dessous de tout, vacillante et sèche, si fragile, si faible!" (*ER*, 590–591).

[59] "J'ai eu toutes les chances; je suis débarqué ici, l'estomac détraqué, le corps malade et les repas laconiques de la Trappe m'ont guéri!" (*ER*, 643).

food, poison or knowledge. Knowledge is inherently problematic, and its assimilation by the human subject is a veritable struggle akin to difficult digestion, in that knowledge encountered by the intellectual constitution has already been produced, digested and repeated, and yet still must be assimilated before being articulated, not least by the literary text.[60]

Huysmans's work too is sensitive to the problematics of predigested discourse, which may seem desirable to the human subject weary of the accumulated excesses of realism and naturalism, to which Des Esseintes is almost a caricature of an over-refined reaction in his paradoxical longing for discursive versions of contemporary mass-produced foodstuffs that ironically enjoy the allure of sublime delicacy. These contradictions are to an extent resolved in *En Route*, which appears to reject dietary refinement, extraction and specificity in both physiological and intellectual realms, blurred in Durtal's experience. The key literary implication of Huysmans's resolution of the conflict between extraction and accumulation is a text that is neither condensed, extracted nor accumulative; *En Route* is a lengthy and exhaustive discussion structured in terms of its thematic and locational shifts into liturgies of the word and of the Eucharist. Ultimately, there are compelling parallels between the protagonist's realisation of the importance of organic integrity and the emergence of a text which abandons both accumulative and extractive models of discursive digestion.

In Zola's case, a similar process occurs over the course of the *Rougon-Macquart* cycle, a byword for the epistemological accumulation of which Zola's predecessor Flaubert is highly sceptical. While Zola's novels discussed here primarily use digestive metaphor to represent accumulative processes, the summative novel in the series acknowledges that accumulation has an implicitly textual dimension and makes a strong case for the primacy of internal organic coherence—both in the body and the text—over the specificity of what either ingest. As in Huysmans, there is a progression towards resolution of the issues raised by the ingestion of knowledge. Both novelists engage from different angles with the accumulationist aesthetic typical of naturalism, before shifting their focus—in both authors' cases, via representations of extraction of both textual and substantial nature—to the integrity of the ingesting organism (body, soul, text). All three authors are attuned to the digestive qualities of the text in their varied representations of digestion.

[60]On the anxiety produced by discursive and alimentary assimilation, see T. A. Unwin, "Eat My Words," 118–129.

Bibliography

Anon. "Un jeu de mot de Flaubert?" Accessed 15 December 2017. http://hircocerf.blogspot.co.uk/2011/01/un-jeu-de-mot-de-flaubert.html.

Bellos, David. "From the Bowels of the Earth: An Essay on *Germinal*." *Forum for Modern Language Studies* 15 (1979): 35–45.

Borie, Jean. *Huysmans: le Diable, le célibataire et Dieu*. Paris: Grasset, 1991.

Brombert, Victor. *The Novels of Flaubert*. Princeton: Princeton University Press, 1966.

Cabanis, Pierre-Jean-Georges. *Rapports du physique et du moral de l'homme [1802]*. Paris: Crapelet, 1805.

Chéron, Jules. *Introduction à l'étude des lois générales de l'hypodermie*. Paris: Société d'éditions scientifiques, 1893.

Dubosquet, L. *Dissertation sur le cauchemar*. Paris: Didot Jeune, 1815.

Duffy, Larry. *Flaubert, Zola and the Incorporation of Disciplinary Knowledge*. Basingstoke: Palgrave, 2015.

Flaubert, Gustave. *Œuvres complètes*. Edited by B. Masson. 2 vols. Paris: Seuil, 1964.

———. *Madame Bovary*. Paris: Flammarion, 1986.

———. Ms. *Dossier génétique* of *Madame Bovary*. Bibliothèque municipale de Rouen, Ms gg 9.

Hamon, Philippe. "Du Savoir dans le texte." *Revue des sciences humaines* 160 (1975): 489–499.

Harrow, Susan. *Zola: The Body Modern*. Oxford: Legenda, 2010.

Huysmans, Joris-Karl. *À Rebours*. Paris: Crés, 1922.

———. *Le Roman de Durtal*. Paris: Bartillat, 1999.

Manzini, Francesco. "Nutrition, Hunger and Fasting: Spiritual and Material Naturalism in Zola and Huysmans." *Forum for Modern Language Studies* 49 (2013): 20–32.

Marquer, Bertrand. *L'Autre siècle de Messer Gaster? Physiologies de l'estomac dans la littérature du XIXe siècle*. Paris: Hermann, 2017.

Mathias, Manon. "Recycling Excrement in Flaubert and Zola." *Forum for Modern Language Studies* 54, no. 2 (April 2018): 224–243.

Otis, Laura. *Organic Memory: History and the Body in the Late Nineteenth and Early Twentieth Centuries*. Lincoln: University of Nebraska Press, 1994.

Petit, H. "Incube." *Dictionnaire des Sciences médicales* 24 (1818): 304–310.

Robert, P. *Le Petit Robert. Dictionnaire alphabétique et analogique de la langue française*. Edited by Alain Rey and Josette Rey-Debove. Paris: Le Robert-SEJER, 2017.

Schaeffle, Albert. *La Quintessence du socialisme*. Paris: Librairie du Progrés, 1880.

Tissot, S. A. D. *L'Onanisme*. Lausanne: Chapuis, 1764.

———. *De la Santé des Gens de lettres*. Lausanne: Dessain, 1775.

Unwin, T. A. "Eat My Words: Verne and Flaubert, or the Anxiety of the Culinary." In *Consuming Cultures: The Arts of the French Table*, edited by John West-Sooby, 118–129. Newark: University of Delaware Press, 2004.

Vigarello, Georges. *Le corps redressé. Histoire d'un pouvoir pédagogique*. Paris: Delarge, 1978.

Zola, Émile. *Les Rougon-Macquart. Histoire naturelle et sociale d'une famille sous le Second Empire*. Edited by H. Mitterand. 5 vols. Paris: Gallimard, 1960–1967.

———. *Germinal*. Paris: Gallimard, 1978.

———. *Nana, dossier préparatoire*, BnF, Nouvelles acquisitions françaises, 10313.

CHAPTER 10

Hygiene, Food and Digestion in Post-Unified Italy: Paolo Mantegazza's Medicine in the Kitchen and Beyond (1861–1900)

Cristiano Turbil

Introduction

The unification of Italy in 1861 created a country where cultural and political differences and traditions had to learn how to cohabit. For hygiene and its popular reception, in particular, the Italian Risorgimento represented a time of challenges but also of political and social significance. One of the key figures who contributed to the establishment of the practice of hygiene in Italy was the polymath Paolo Mantegazza. In just over thirty years of constant work Mantegazza contributed, like no other, to the establishment of popular medicine in Italy. His efforts to make hygiene popular were conducted through initiatives on both a professional and popular level. Mantegazza promoted hygiene in a variety of ways: from laboratory medicine and sex education to food science and digestion. A study of Mantegazza's works on food tells part of the

C. Turbil (✉)
Historian of Science and Medicine, King's College London, London, UK

M. Mathias and A. M. Moore (eds.), *Gut Feeling and Digestive Health in Nineteenth-Century Literature, History and Culture*, Palgrave Studies in Literature, Science and Medicine,
https://doi.org/10.1007/978-3-030-01857-3_10

history of Italian medicine. More specifically, it can show how the study of hygiene did not only contribute to the extension of scientific literacy but also to constructing a sense of national identity and cohesion in a country which had long been divided.

Before looking at how Mantegazza popularised and discussed food and digestion, it is important to look briefly at the place of Mantegazza's work in contemporary critical scholarship. In the current scholarship on the Italian Risorgimento, Mantegazza does not occupy a prominent place. Only a few studies have been published in the last twenty years focusing on very particular aspects of his work. In Anglo-American scholarship, Mantegazza's work has been recently rediscovered by historians and literary scholars. A translation of his scientific and literary works has been produced recently by Nicoletta Pireddu (2007).[1] Mantegazza has also been recognised as a pioneer of sexual medicine in the nineteenth century and the majority of studies focus on this particular aspect of his work.[2] However, little has been written on Mantegazza's work on food science and digestion and the role they played in improving Italian public health. In Italian academia, specific research on Mantegazza is also limited to a very small number of scholars. Mantegazza is again studied mostly for his contributions to the popular history of sexuality.[3] Mantegazza's fame, as clearly explained by Claudio De Boni, is in no way close to what it was in the late nineteenth century.[4]

During his life, Mantegazza was well known not only in Italy but all across the Western world. By the 1880s, he was recognised as a leading international scientist and was cited by important scientific figures including Sigmund Freud and Charles Darwin.[5] In August 1893, Frederick Starr, publishing a sketch on Mantegazza in the American magazine *Popular Science Monthly*, compared Mantegazza to Pierre

[1] Mantegazza, *The Physiology of Love.*

[2] See De Feo, "Paolo Mantegazza as Didatic Gastronome," 26; Dell' Olio, "Paolo Mantegazza: memoria sulle proprietà terapeutiche della coca"; Garbarino and Mazzarello, "A Strange Horn Between Paolo Mantegazza and Charles Darwin"; and Pireddu, "Between Darwin and San Francesco: Zoographic Ambivalences in Paolo Mantegazza, Ouida, and Vernon Lee."

[3] See Moruno, "Love in the Times of Darwinism" and "Pain as Practice in Paolo Mantegazza's Science of Emotions"; Ross, "Pathologies and Eroticism: Paolo Mantegazza's Ambiguous Reflections on Female Same-Sex Sexuality."

[4] De Boni, "Utopia e positivismo: il caso italiano di Paolo Mantegazza."

[5] Dell' Olio, "Paolo Mantegazza: memoria sulle proprietà terapeutiche della coca."

Broca, the father of French anthropology, saying: 'What Broca was to Paris and then to France, Mantegazza is to Italy'.[6] Starr's sketch suggested that Mantegazza's scientific and popular books were read all across the Western world.[7]

This chapter will look at two specific aspects of Mantegazza's writing on hygiene focusing in particular on his work on food and digestion. It will rely on three different types of primary sources authored by Mantegazza: a professional medical volume *Elementi D'Igiene* (Elements of Hygiene, 1864), a popular series of books called 'Almanacchi Igienico popolari' (*The Popular Hygenic Almanacs*) and a selection of passages from Mantegazza's fictional works. My claim is that Mantegazza considered digestion as a political matter because he recognised that a society that eats and digests well is a modern and hygienic society. In this respect, an analysis of the political function played by the stomach in the establishment of post-unified Italian public health will show how food and digestion became an aspect of primary concern for the Italian medical and political community. Beside the analysis of the context, which will be central in the exploration of the science of digestion in the period, this chapter will also offer an opportunity to look into how Mantegazza's language, terminology and style of writing changed across different platforms.

Introducing the 'Alimenti Nervosi'

In order to shed some light on the medical, social and cultural importance of Mantegazza's work on hygiene, food and digestion, one has to start from two key aspects: on the one hand, his own research, and on the other hand, his ideas regarding the political importance of popularising medicine. Those two aspects, as we will see, were intrinsically connected to every activity undertaken by the Italian doctor.

Medically trained in both Pisa and Pavia, Mantegazza, like many other young men of his generation, spent a long period travelling around the world for research, for training and for pleasure. He travelled first to Germany, France and England where he directly observed the introduction of revolutionary new public health campaigns, and then to

[6] Starr, "Sketch of Paolo Mantegazza."

[7] Ibid., 550. See also Robinson, "Introduction," in *The Sexual Relations of Mankind*, edited by Mantegazza, ix.

South America where he lived for several years working as a surgeon in Argentina, Paraguay and Bolivia.[8] The South American period provided Mantegazza with the opportunity to encounter different cultures and ideas but also to conduct innovative anthropological and medical research which would significantly influence his future work.

Returning to Italy in 1859, he first worked as a surgeon in Milan. As a young scientist interested in the development of the discipline, Mantegazza clearly understood the new direction science was taking in other European countries. The place of medicine was not limited only to a university, hospital or a laboratory; it was also located in the public square, among common people. It is within this context that Mantegazza started promoting his research and ideas concerning food and hygiene.

In 1859, Mantegazza published a paper on the outcomes of his South American research on the medical use of coca leaves entitled 'Sulle virtù igieniche e medicinali della coca e sugli alimenti nervosi in generale' ('On the hygienic and medicinal properties of coca and on nervine nourishment in general') in the periodical *Annuali Generali di Medicina* (Annuals of General Medicine). The paper had a specific aim. Mantegazza presented a medical overview of the main 'alimenti nervosi' (foods that stimulate the nervous system) and advanced the conclusion that a measured consumption of those nourishments should become part of everyday diet.[9] Mantegazza asserted the importance of consuming coffee, coca leaves or various alcoholic distillates to obtain certain medical benefits for particular types of diseases or medical conditions.[10] This was the first of several articles that Mantegazza wrote investigating the importance of certain foods on digestion, the body and consequently more generally public health. Indeed, reading this paper, it is possible already to delineate some of the key concerns that became central to his later work. In discussing the medical importance of consuming various 'alimenti nervosi', Mantegazza immediately established a clear link between the digestion of those foods and the effect they have on the neurological system. In particular, he explained how the stomach and

[8]Govoni, *Un pubblico per la scienza*, 110–111.

[9]Mantegazza, "Sulle virtù igieniche e medicinali della coca e sugli alimenti nervosi in generale."

[10]Mantegazza, "Sulle virtù igieniche e medicinali della coca e sugli alimenti nervosi in generale," 473–485.

the brain were directly linked, with the consequent ability to control and stimulate the latter by simply digesting certain foods.

In this respect, Mantegazza's specific study on coca leaves is a perfect example to understand the importance that the stomach played in both the physiological and psychological well-being of the individual. In enumerating the various benefits that a correct consumption of coca leaves provides, Mantegazza observed that coca could be used as a support food for both physical strength and the spirit. The article provides guidelines for obtaining certain benefits by chewing a specific amount of coca but also more generally an explanation of how the correct eating of this plant could stimulate the brain and enhance certain physical and intellectual abilities. As indicated in the text, this 'alimento nervoso' could be used for curing gastrointestinal diseases and for whitening and improving the health of both teeth and gums.[11] However, the beneficial effects of coca leaves were not limited to the health of the body. Indeed, the doctor explained that an infusion of this plant could be useful to help digestion after a rich lunch, keeping the brain active and helping concentration.

Mantegazza provided reports of two direct observations of the effect that coca has on the body. First, observing the locals, the Italian doctor showed how chewing a specific amount of coca could allow an individual to work for longer hours without eating or sleeping.[12] Second, testing coca directly on himself, he explained that a measured consumption of coca leaves could boost the mental abilities of the individual especially allowing him to explore complex theoretical thoughts.[13]

In 1859, Mantegazza published another essay on coca entitled 'Sull'introduzione in Europa della coca, nuovo elemento nervosa' (On the introduction of coca in Europe as new nervine element) in the medical journal *Annali di Chimica applicata alla Medicina* (Annuals of Chemistry Applied to Medicine). This second essay had a different aim. More focused in scope, it presented an analysis of the chemical composition of the coca, explaining how it was possible to isolate from the plant

[11] Ibid., 506–508.

[12] Ibid., 490–492.

[13] Ibid., 493–494.

a powerful but also very practical anaesthetic to be used in surgery.[14] As observed by the historian Giuliano Dell' Olio, this publication contributed to making Mantegazza internationally famous.[15] However, it also contributed to further clarifying the growing importance of medical research of the 'alimenti nervosi' and the incredible possibility they offered for both medicine and society. The study of the 'alimenti nervosi' is, therefore, intrinsically linked to Mantegazza's medical, political and cultural work. References to these foods can be found in all of Mantegazza's writing from the political essays to the fictional writings and from academic texts to popular articles.

As with other leading Italian figures of the period, including the philosopher Roberto Ardigò and the father of criminology Cesare Lombroso, Mantegazza advocated the teaching of positivism. Especially in his medical and anthropological research, he recognised the need to consider scientific knowledge as the final and most important type of knowledge even beyond the humanities and politics. As suggested by the historian Claudio De Boni, Mantegazza's opinion was influenced by the work of Pierre Broca (1824–1880) who saw in anthropology the perfect link between science and history.[16] Indeed, for Mantegazza, the study of anthropology was not simply a question of looking at the history of one culture or population. It had the positivist spirit of helping the development of the society of the future. In this respect, Mantegazza's work on hygiene and digestion followed the same path.

Indeed, a key aspect of his publications was the necessity to teach both the medical community and the general public alike. In this respect, it is important to highlight how the study of food and digestion had a specific political meaning in Mantegazza's work. In the next few pages, I will analyse several passages from both Mantegazza's medical and popular writing on food and digestion to show how he mixed science with politics in order to guide his readers, even on a moral level.

[14] Mantegazza, "Sull'introduzione in Europa della coca, nuovo elemento nervoso," 18–21.

[15] Dell' Olio, "Paolo Mantegazza: memoria sulle proprietà terapeutiche della coca," 228–239.

[16] De Boni, "Utopia e positivismo: il caso italiano di Paolo Mantegazza," 138.

Food, Politics and the 'Alimenti Nervosi'

In order to explore the relationship between food, digestion and Italian politics one has to start from the *Elementi di Igiene*, one of Mantegazza's most famous and well-known books. The volume, published originally in 1864 and then updated and reprinted several times, has a double value for historians. First, it provides a clear map of Mantegazza's scientific understanding of hygiene in its practical and theoretical forms. Second, the volume provides an overview of how hygiene developed across the Western world over the nineteenth century. The volume, since its first edition, included a section called 'Biblioteca dell'Igenista' (Library of the Hygienist) where Mantegazza reported, in thematic sections, a bibliography of the main studies concerning any aspect of hygiene published in Europe since the beginning of the century. Mantegazza considered this section as a key element of the volume and, in each new edition, he always tried to keep his study up to date with new cutting-edge research and reports from all around the continent.

In the first part of the volume entitled 'analytic hygiene', Mantegazza scientifically explores several aspects of what he calls the 'human machines'. This part of the volume, more medically oriented, provides a physiological analysis of how to improve health throughout the implementation of specific hygienic norms with the aim of creating a culture of well-being. Mantegazza's aim was to write a medical guide to explain, not only to the medical community but also to a wide lay readership, the importance of hygiene for both the individual and the state. The study of food and digestion, he claimed, was central to understanding the human body. Indeed, one of the key concepts advanced by the doctor at the beginning of this volume was that a good level of hygiene in the kitchen combined with medical knowledge of food can provide 'the precious gift of good digestion'.[17] This 'precious gift' was not only beneficial to the single individual but also to the community as a whole.

In the *Elementi di Igiene*, Mantegazza conceived and described the life of the individual and the species as two separate macro-categories each with its own goals and duties. For the Italian doctor, each individual has two main objectives in life: to feed and to feel or experience

[17] "Un uomo sano potrà imparare a digerir bene, leggendo queste pagine, e continuerà a conservare per tutta la vita il prezioso tesoro d'una buona digestione, se ubbidirà a tutti gli altri precetti d'igiene" (Mantegazza, *Elementi d'Igiene*, 66).

the world.[18] Within the first objective lay the necessity to introduce correct nourishment to the body (and consequently digest it) in order to allow the correct physiological functions of the organism. Those functions included: the circulation of blood throughout the body, the ability to breathe and the need to digest and defecate the residue of the digestive process.[19] Within the second objective, feelings, Mantegazza included the five senses and the ability to move, speak and think. The second macro-category, the life of the species, predominately related Mantegazza's hygienic concerns about human reproduction and the improvement of the next generation.

From the beginning, the author made clear the need to medically study food and digestion in relation to the new discoveries made by the science of hygiene. The goal was to provide an overview of the whole process from the mastication of food, through its digestion, concluding with its defecation. This analysis was supported by a series of practical recommendations concerning the health of the body parts involved in the eating process. For instance, Mantegazza provided several guides, supported by empirical experiments, on how to cure diseases inside the mouth, on how to improve the health of the teeth, avoiding stomach ache and other known diseases of both the stomach and the intestine. Nonetheless, the importance of hygiene went beyond the body. Mantegazza highlighted the risks concerning the spread of diseases created by communal eating, by the sharing of cups or other tools that come into direct contact with the mouth.

In the *Elementi*, Mantegazza also examined all of the possible foods that can be found on the table of any Italian family: from the kitchen of the rich and educated to the table of the average working-class household. Mantegazza divided nourishments into three macro-categories: *Plastici, Combustibili and Nervosi* (Plastic, Combustible and Nervine).[20] For each of those categories, he provided a detailed analysis of their chemical properties and a guide on how to cook and eat them in order for our body to make the most of their nutrients. It is here that the professional and international breadth of the volume becomes clear. In composing those sections, Mantegazza did not only rely on his own

[18] Mantegazza, *Elementi d'Igiene*, 21.

[19] Ibid., 21.

[20] Ibid., 58.

research but also cited and discussed the work of numerous European colleagues. As stated earlier, the analytical study of foods in the *Elementi* was designed to be used as guidance for Italian doctors.[21]

There are several case studies presented in the volume that are worthy of close study. However, one in particular shows how Mantegazza's medical research was deeply linked to his political vision. In Chapter 4, he discussed the medical benefit of drinking water and consuming the right amount of salt in everyday diet. In the text, Mantegazza is particularly keen to explain the effect that water and salt could have on the physiological functions of the body. After a detailed analysis of the different types of water that might be drunk in different parts of the globe and the effect they have on the organism, the author ventures into explaining the importance of salt for the physiological well-being of the human body. Before explaining why it is important to include salt in the everyday diet, Mantegazza explained the difference that a diet which includes salt (*alimentazione salata*) and one without it (*alimentazione senza sale*) has on the chemical composition of human blood, and consequently on the general health of the individual.[22]

Mantegazza recognised primarily the necessity to eat correctly according to the physiological need of the body. He explained that the healthy man should not eat excessively while a sick individual should always follow the advice of a doctor and limit himself even if he struggles with a compelling urge due to his disease.[23] The same can be said about drinking. Indeed, he recognises that feeling angry or thirsty must be considered as the two main physiological alarms used by the body to communicate to us. Mantegazza insisted that 'anytime they call, we need to reply immediately: I am here'.[24] But the implications of ignoring such

[21] Mantegazza, "Economia gastronomica – Igiene della cucina," 12–14; Mantegazza, "Degli alimenti nervosi narcotici – Il tabacco – La coca - L'oppio," 23–24.

[22] Ibid., 80.

[23] "L'uomo sano che ha poco fame deve mangiar poco; l'uomo sano che non ha fame non deve mangiare. L'uomo malato che non ha fame non deve mangiare; l'uomo malato che ha fame deve consultare il medico, onde lo aiuti ad interpretare la natura di questo bisogno; anche quando la sua fame fosse violentissima, insoffribile" (ibid., 32).

[24] "L'interpretazione della sete dev'esser fatta come quella della fame. Sono due sentinelle benevole che ci avvertono dei nostri bisogni, e così come non vogliono essere eccitate a dare falsi allarmi, così non si può lasciarle gridar troppo, a rischio che perdano la voce e non ci avvertano più colla consueta fedeltà del vicino pericolo. Al primo chi va là della fame e della sete conviene subito rispondere: presente" (ibid., 32).

needs extended beyond the health of the individual, into the vitality of the nation.

In the second part of the volume, Mantegazza presented to his readers what he labelled a 'philosophy of hygiene'. The scope, here, far from being a medical examination of the human body, was to shed some light on the social and political importance of health. On the matter of eating and digestion, the doctor is very specific: Mantegazza insisted on the necessity, for any nation, to invest in studying the 'physiology of nutrition'—a matter, in his opinion, still very obscure in Italy at this time.[25] This, he explained a few pages later, was because the physical and intellectual capabilities of individuals determined the physiology of the nation as a whole.[26] The analogy between the body and the nation is further explained in what Mantegazza called the 'economy of vital forces'.[27] This physiological economy, as with the real economy, needs to be extremely precise in order to work properly. Indeed, the Italian hygienist claimed the need for a political revolution that should make the introduction of hygienic norms a key priority of any modern state.[28] In this respect, 'the precious gift of a good digestion' had to be considered of primary political importance.

Not surprisingly these ideas had a clear impact on his political activity. In 1865, Mantegazza was elected, for the municipality of Monza, a member of the Italian parliament. He served first in the parliament and then in the senate for over twenty years. His main political ambition was, as suggested by the historian Giuliano Pancaldi, to create and consolidate a new secular state based on his ideas about social Darwinism.[29] Walter Pasini explains that Mantegazza saw politics as an extension of his medical activity.[30] Therefore, for Mantegazza, politics was the only way to implement his medical ideas and vision of science on a national level. In this respect, the study of food and digestion became more than a medical investigation. It became a tool to be used by both politicians and the medical community to improve the quality of life of the whole population.

[25] Mantegazza, *Elementi d'Igiene*, 566.

[26] Ibid., 573–574.

[27] Ibid., 578–580.

[28] Ibid., 573–574.

[29] Pancaldi, *Darwin in Italia. Impresa scientifica e frontiere culturali*, 15.

[30] Pasini, *Paolo Mantegazza ovvero L'elogio del eclettismo*, 122.

In 1896, Mantegazza published his political memoir *Ricordi Politici di un Fantaccino del Parlamento Italiano* (Political Memoirs of a Foot Soldier in the Italian Parliament).[31] The volume provides an overview of both his political ideas and his experiences in the Italian parliament. One episode is particularly relevant here. On the 15th of May 1870, the Italian Minister of Finance, Quintino Sella, commissioned a study into the cost and usage of salt in the baking of bread across Italy.[32] The study was designed to provide information towards new legislation on the price of cereals soon to be discussed by the Italian parliament. As reported in the volume, Sella appointed Mantegazza for two main reasons: his medical expertise and the political significance of his research.

The Italian doctor recognised that previous studies on the matter lacked any scientific methodology, did not encompass the recent studies on hygiene and, finally, did not take into consideration the different Italian regions and their traditions. For Mantegazza, the question of salt was not just related to the consumption of food but was rather 'a question of hygiene and political economy'.[33] Indeed, Mantegazza's report presents scientific knowledge over political decisions while also making clear how any commission should focus their effort on understanding the physiological need of the body rather than simply focusing on the cost of a product and the consequence this has on the economy. This point was not new to Mantegazza. Indeed, the question had already been explored by the author in an article published in 1866 in the periodical *L'igea* (Hygiene) where he discussed the parliamentary debate about salt and taxation again promoting the same idea.[34] Mantegazza's need to make science political became the main theme of several volumes of another important series of books published by this author during his career, the *Almanacchi Igenico Popolari* (*The Popular Hygenic Almanacs*). Published for over 40 years, they were one of the most successful series of medical vulgarisation in Italy in the first decades after the unification.[35] One hundred and fifty pages long, the almanacs had the aim of teaching Italians basic medical ideas to make them a 'doctor at home'.

[31] Mantegazza, *Ricordi Politici di un Fantaccino del Parlamento Italiano*, 18–19, 56–58.

[32] Ibid., 155.

[33] Ibid.

[34] Mantegazza, "L'ultima legge sul sale alla Camera dei Deputati," 209–211.

[35] Govoni, *Un pubblico per la scienza*, 132–139.

Recent historical scholarship has explored the meaning and importance of popular science, especially in Victorian England.[36] The popularisation of science was a process so common as to convince contemporary historians such as Lightman and Aileen Fyfe to use the analogy of the marketplace.[37] In Italy, however, the situation was rather different and therefore the analogy of the marketplace had to be reformulated in relation to a different set of social and historical conditions. Paola Govoni, in her recent work on Italian popular science, explains that in 1870s Italy, the phenomenon of popularisation was rather recent because of the complex political climate.[38] Italy was not a united country until 1861 and, therefore, the spread of scientific knowledge and education was very different across the peninsula. In Italy in 1871, Govoni explains that over 60% of the population was unable to read and write. In addition, there was a clear discrepancy in education between the north (54% illiterate) and the south (84% illiterate) of the country.[39]

In the light of this, Mantegazza's almanacs can be easily defined as an editorial success in the young Italian kingdom. Published by the editor Treves of Milan, a specialist in affordable popular science, the almanacs were designed around a simple concept. They were not created to be academic or professional texts like the *Elementi*. Instead, the aim was to make science practical. Although every single almanac had a specific topic, the whole series had a common theme: once again public health. Mantegazza provided precise guides on how to improve hygiene at home or at work, on how to preserve food, on how to use different types of soaps, how to cook certain vegetables or types of proteins or even how to use medications. Besides the practical aspect, the other aim of the almanacs was, again, political. Mantegazza saw the need to teach and show the importance of science and medicine to any citizen, not only those with an education.

There are several almanacs focused on topics of food and digestion. In 1872, Mantegazza published *L'Igiene D'Epicuro* (Hygiene of Epicurus). The aim of the almanac appeared clear from the preface. Mantegazza's intention was to return to the topics explored in

[36] See Topham, "Introduction Focus: Historicizing 'Popular Science'"; Secord, "Knowledge in Transit."

[37] Lightman and Fyfe, eds., *Science in the Marketplace*, 1–23.

[38] Ibid., 56–58.

[39] Ibid., 109.

the *Elementi*, but this time, without providing the same level of detail as the main scientific publications. The tone, language and examples used in the almanacs are designed to be easily understood and replicated by everyone at home. The *Igiene D'Epicuro* focused particularly on Mantegazza's research on the 'alimenti nervosi' as the author considered it particularly important to give the general public a clear understanding of these nourishments. At the very beginning of the volume, Mantegazza made clear, once again, the importance that food has for the organism, using the analogy of the human body as a machine that needs a specific fuel to work properly.[40] However, attention was also devoted to the brain and the intellectual abilities of the individual.[41] In this respect, a precise explanation of the properties of the 'alimenti nervosi' became necessary. The volume explained that when the brain does not think, work or sleep, it suffers. The brain has, therefore, to be stimulated by an 'alimento nervoso' like coffee, tobacco or wine. Mantegazza made, of course, a clear distinction between individuals and their mental capabilities but his main hypothesis remained the same. The doctor suggested the need to introduce the 'alimenti nervosi' into the Italian diet but also to educate people about their use.[42]

If compared to the *Elementi*, the main difference between the two texts is in the tone, language and complexity of the explanation. As with the *Elementi D' Igiene*, the almanac also included a small number of tables. However, the tables in the almanac are extremely simplified and only designed to provide data regarding affordable ali*menti nervose*—spirits in particular. This is because the scope of the almanac was not to be analytical or to teach the reader about science and medicine but, instead, to educate and provide moral guidance. This became clear in the final chapter where Mantegazza points out the moral of the volume. He uses the examples of a farmer and explains how his passion for drinking jeopardised his opportunities of having a good and wealthy life. The author defines drunkenness as the moral death of the self, which meant 'the transformation of a man into an animal'.[43] This episode serves, in the economy of the volume, to explain how consuming alcohol (or other

[40] Mantegazza, *Almanacco Igienico*, 10

[41] Ibid., 13.

[42] Ibid., 12.

[43] Mantegazza, *Almanacco Igienico Popolare Igiene della casa*, 131

alimenti nervosi) although useful in certain cases could be extremely dangerous for the individual (and the wider community) if abused.

A similar argument was already advanced a few years before in 1867 in the almanac *Igiene della casa* (Domestic hygiene) where Mantegazza explained how it was possible to understand the character of an individual simply by looking at the level of hygiene in his home. In the almanacs, Mantegazza transferred in a popular form some of his research concerning the introduction of new hygienic norms inside the kitchen and bathroom published in 1862–1863 in *L'igea*. In those articles, he explained to the medical community the importance of keeping certain areas like the kitchen and bathroom clean and the effect this would have on health.[44] For the author, an honest person can live only in a clean and well-organised house whereas a dirty and smelly household can only belong to a criminal.[45] The same point is expanded later on in the volume where Mantegazza, looking at the household from a moral standpoint, writes: 'The house is a moral school and populations are as much honest as they love, look after and make beautiful their homes. The vicious man, not different from a ferocious beast, does not get home until late in the night and only to sleep; the honest man, instead, loves his house and look after it. There, he spends happy and quiet hours with his family and friends'.[46]

Further evidence regarding Mantegazza's political use of food is enclosed in 'Dizionario della cucina' (Dictionary of the kitchen), an almanac published in 1882. In this volume, Mantegazza presented in alphabetical order all the possible dishes that can be found in Italy. However, once again, Mantegazza's catalogue was not just an overview of the main Italian foods: the volume provides, instead, a list of anecdotal information about Italian ingredients and detailed guides on how and when to eat them. In this volume, the author explains and compares

[44] Mantegazza, "Badate al cesso," I: 466–478.

[45] "Tale è l'uomo e tale è la casa: la capanna spiega il selvaggio, la catapecchia descrive il contadino, il palazzo vi racconta il principe. La casa ordinata e pulita vi dice l'uomo onesto, l'uomo felice: vi dice salute, economia, eterna serenità di affetti. La casa sudicia e fetente vi dice scrofola e vizio; vi dice bestemmie e busse; fame e vituperio.' Igiene della casa" (Mantegazza, *Almanacco Igienico*, 9).

[46] "La casa è scuola di morale, e i popoli sono tanto più onesti quanto più l'amano e la coltivano e la fanno bella. L'uomo vizioso, poco diverso dalle fiere, non ritorna alla casa che a tarda ora e per dormirvi. L'uomo onesto vi lavora e vi ama; vi passa ore liete e serene cogli amici o i figliuoli" (ibid., 13).

traditional dishes from the north and the south of the country to show how even different traditions had something in common. Once again, here, Mantegazza writing on food has a specific political aim: the need to forge a sense of national identity. In the preface of the volume, he explained that eating is more than a physiological need. It is a ritual that brings happiness to both the individual and the household (and consequently to the community).[47]

However, the knowledge of different foods also had an important cultural value. The author highlighted the link between different products and their local traditions. For instance, with anchovies, besides their culinary properties and some advice on how to cook them, the doctor explained this product's importance not only for the local but also the national economy.[48] Another example comes from the section about coffee. In this case, instead of explaining the properties of this beverage (the topic already largely explored in his work on the 'alimenti nervosi'), he compared the preparation of coffee to the work of the genius. The doctor explained that the preparation of this beverage needs two specific skills, precision and timing, two things that according to Mantegazza should be key to any good political work.[49] The reference to politics, here, although limited, still fits with his plan to promote a correct use of this 'alimento nervoso'. In order for any individual to enhance his mental abilities (almost to the level of a politician), a correct preparation and consumption of coffee is therefore necessary.

Mantegazza's intention to educate the Italian population went far beyond the writing of the almanacs. In 1868, he recognised that the 'lower classes' were not able to grasp the complexity of his scientific essays. Therefore, as explained in the preface of his most successful fictional work, *Un Giorno a Madera* (A Day in Madera, 1868), Mantegazza started writing novels without pretending to be a great novelist but with the desire only to write a 'useful and moral book'.[50] Literature, then, became a tool to allow the 'lower classes' to benefit from science.

In the light of this, it is interesting to look at how the doctor defined himself in 1909, in his final book, *La Bibbia della Speranza* (*The Bible*

[47] Mantegazza, *Almanacco Igienico*, 8–9.

[48] Ibid., 11–12.

[49] Ibid., 30.

[50] Mantegazza, *Un giorno a madera*, 8.

of Hope), as neither an important scientist nor a novelist. He instead described himself as a bit of both: 'a polymath of many intellectual interests'.[51] Although not the main aim of Mantegazza's fictional work, references to food and digestion are varied and always used in such a way to promote the author's ideas as widely as possible. Literature was considered by the doctor as an opportunity to put his ideas on accessible display for the general public.

In the novel *Testa: Libro Per I Giovinetti* (*Testa: A Book for Boys*, 1887) Mantegazza, while promoting the learning of science to school-age children, explains the importance of a healthy diet. For Mantegazza, it was particularly important to eat the right amount of food at the right time to help boys in their studies. In the volume, he promoted the eating of certain vegetables[52]; the importance of not eating with greediness to avoid stomach ache and bad dreams, and finally the importance of eating certain types of food according to your age and career.[53] Again a link between the stomach and the brain is explored but this time in a more political manner. Indeed, the novel presents to young Italians Mantegazza's desire concerning the creation of national science, the topic of which the doctor made himself political ambassador a few years before. In an 1880 talk on 'the new Italian science' (La scienza nell'italia Nuova) and in an 1884 article published in *La Natura*, Mantegazza stated the need for the creation of a national science.[54] He explained that the improvement of any scientific knowledge should be the highest aspiration of any civil society: 'we should not have a science of the right or of the left, there should be a science of the nation'.[55]

Mantegazza continued to promote his ideas regarding food and digestion and their political importance in *L'Anno 3000* (*The Year 3000*), his last fictional work. Although direct reference to food and digestion are, limited here, the novel is the perfect platform to see how all of his political, social and scientific ideas come together. *The Year 3000* combines Mantegazza's interest in science with his passion for politics, allowing the author to carefully discuss and popularise the combination of the two.

[51] "Poligamo di molti amori intellettuali" (Mantegazza, *La Bibbia della Speranza*, 112).

[52] Mantegazza, *Testa: A Book for Boys*, 45.

[53] Ibid., 50.

[54] Turbil, "Paolo Mantegazza and the Dream of 'Making' Science Popular," 1–5.

[55] Mantegazza, *La Scienza nell'Italia Nuova*, 18.

In the novel, Mantegazza makes explicit, in a language suited to the working class, ideas already discussed in his scientific and popular work.

The Year 3000 is the story of a journey into the future (a dream as indicated in the subtitle), where Paolo and Maria, the two protagonists, travel from Rome to Andropolis, the capital of the United planetary state, at the foot of the Himalayas, to ask permission to get married and procreate. In the fictional future described by Mantegazza, procreation is a global matter that has to be centrally regulated in order to prevent overpopulation.

The Year 3000 is as observed by Nicoletta Pireddu, one of the first examples of Science Fiction in Italy. However, in line with all of his other publications, Mantegazza's concerns were very much about the present. In fact, it is important to see Mantegazza's literature as a necessary continuation of his scientific and political work instead of a simple way of experimenting with different genres. In *The Year 3000*, he explains that human society should be subject to the rule of science and not governed by ideas like socialism:

> I believe that the speed of communications, obtained with steam and telegraph, has contributed more than all the books, all the newspapers, all the parliaments, laws, and even all religions; to destroy the criminal old order of war among peoples and to create a new moral, sane and sincere.[56]

Mantegazza's political views become even clearer in Chapter 5 when Paolo and Maria visit the government palace in the capital of the world. The palace is divided into four sections, each of them dedicated to an important scientific area: land, health, school and industry. In Mantegazza's fiction, the state is not designed to be governed by elected members. Everything in Andropolis is organised according to nature and run by hygienists. In this fictional world, there is no war and everyone lives a happy and healthy life. This is because the society of the future fully embraces all of the ideas that Mantegazza advanced in both his medical and political work.

[56]"Io credo che la rapidità delle comunicazioni, - continuava a dire l'ingegnere, - ottenute col vapore e col telegrafo hanno contribuito più di tutti i libri, di tutti i giornali, più di tutti i parlamenti, di tutti i codici ed anche di tutte le religioni a distruggere l'antica e scellerata epoca delle guerre fra popolo e popolo e a creare una nuova morale; sana e sincera" (Mantegazza, *L'Anno 3000. Sogno*, 65).

In conclusion, the study of food and digestion in the context of Italian unification, and more broadly within the medical revolution produced by hygiene, shows how the stomach became more than an organ designed to digest. It became a medical but also a political tool—to be used both by the individual and also by the political class in the interest of the nation.

In this respect, it is no surprise to read Mantegazza's work on digestion as a critical overview of the main issues of his own century. In his thirty-year career as both politician and hygienist, Mantegazza proposed a clear analogy between the young Italian nation and the body of a child that was slowly growing into adulthood. Keeping this analogy in mind, the Italian doctor promoted a revolution in terms of both the 'physiology of nutrition' and the 'economy of vital forces'. Partially medical and partially political, this plan for the future aimed to change how Italians treated their own bodies in order to create a new, modern, moral and healthy Italian nation.

Unfortunately, this plan did not succeed as much as Mantegazza had hoped. Nonetheless, his work indicates three different yet related concepts for further historical investigation about medicine in post-unified Italy: firstly, the need to treat the study of food in the nineteenth century as a hygienic concept rather than simply an economic factor. Secondly, Mantegazza's study of special types of food like the 'alimenti nervosi' developed the link between the stomach and the mental capacity of the individual. Thirdly and finally, Mantegazza's writing shows that, in post-unified Italy, there was a clear continuity between political power, medical research and its public understanding.

Bibliography

Chiarelli, Cosimo, and Walter Pasini, eds. *Paolo Mantegazza: Medico, Antropologo, Viaggiatore*. Firenze: Firenze University Press, 2000.

De Boni, Claudio. "Utopia e positivismo: il caso italiano di Paolo Mantegazza." *MORUS* 4 (2007): 137–148.

De Feo, Daniele. "Paolo Mantegazza as Didatic Gastronome: Food, Art, Science and the New Italian Nation." *Humanities* 5 (2016): 26.

Dell' Olio, Giuliano. "Paolo Mantegazza: memoria sulle proprietà terapeutiche della coca." *La Rivista Italiana della Medicina di Laboratorio—Italian Journal of Laboratory Medicine* 7, no. 4 (2011): 228–239.

Garbarino, C., and Paolo Mazzarello. "A Strange Horn Between Paolo Mantegazza and Charles Darwin." *Endeavour* 37, no. 3 (2013): 184–187.

Govoni, Paola. *Un pubblico per la scienza. La divulgazione scientifica nell'Italia in formazione*. Roma: Carocci Editore, 2002.

Lightman, Bernard, and Aileen Fyfe, eds. *Science in the Marketplace: Nineteenth-Century Sites and Experiences*. Chicago: University of Chicago Press, 2007.

Mantegazza, Paolo. "Sulle virtù igieniche e medicinali della coca e sugli alimenti nervosi in generale." *Annuali Generali di Medicina* 167, no. 501 (1859a): 449–519.

———. "Sull'introduzione in Europa della coca, nuovo elemento nervoso." *Annali di Chimica applicata alla Medicina* 29 (1859b): 18–21.

———. "Badate al cesso." *L'Igea, Giornale d'igiene e medicina preventiva* I (1863): 466–478.

———. "Degli alimenti nervosi narcotici – Il tabacco – La coca - L'oppio." *L'Igea, Giornale d'igiene e medicina preventiva* II (1864a): 23–24.

———. "Economia gastronomica – Igiene della cucina." *L'Igea, Giornale d'igiene e medicina preventiva* II (1864b): 12–14.

———. "L'ultima legge sul sale alla Camera dei Deputati." *L'Igea. Giornale d'igiene e medicina preventiva* IV (1866): 209–211.

———. "Carlo Darwin e il suo nuovo libro." *Nuova Antologia* 8 (1868): 1–37. Firenze.

———. *Almanacco Igienico Popolare Igiene D'Epicuro*. Milano: Gaetano Brigola Editore, 1872.

———. *Elementi d'Igiene*. Milano: Gaetano Brigola Editore, 1874.

———. *Fisiologia dell'Amore*. Milano: Brigola Editore, 1879.

———. *La Scienza nell'Italia Nuova. Annuario del R. Istituto di Studi superiori pratici e di perfezionamento*. Firenze: Le Monnier, 1881.

———. *Almanacco Igienico Popolare Dizionario della cucina*. Milano: Gaetano Brigola Editore, 1882.

———. *Testa: A Book for Boys*. Lexington: D.C. Heath, 1888.

———. *Memorie Politiche di un Fantaccino del Parlamento Italiano*. Firenze: P. Bemporand & Figlio, 1896.

———. *La Bibbia della Speranza*. Torino: Sten, 1909.

———. *Un giorno a madera. Una pagina dell'igiene d'amore*. Milano: Casa Editrice Bietti, 1925.

———. *L' Anno 3000. Sogno*. Milano: Fratelli Treves, 1987.

———. *The Physiology of Love and Other Writings*. Toronto: University of Toronto Press, 2007.

Moruno, Dolores. "Love in the Times of Darwinism: Paolo Mantegazza and the Emergence of Sexuality." *Medicina & Storia* 19–20 (2010): 147–164.

———. "Pain as Practice in Paolo Mantegazza's Science of Emotions." *Osiris* 31, no. 1 (2016): 137–162.

Pancaldi, Giuliano. *Darwin in Italia. Impresa scientifica e frontiere culturali*. Bologna: II Mulino, 1983.

Pasini, Walter. *Paolo Mantegazza ovvero L'elogio del eclettismo*. Rimini: Panuzzi, 1999.

———. "Mantegazza Politico." In *Paolo Mantegazza: Medico, Antropologo, Viaggiatore,* edited by Cosimo Chiarelli and Walter Pasini, 121–132. Firenze: Firenze University Press, 2002.

Pireddu, Nicoletta. "Paolo Mantegazza, Fabulator of the Future." In *The Year 3000. A Dream*, edited by Paolo Mantegazza, 1–53. Lincoln: University of Nebraska Press, 2003.

———. "Between Darwin and San Francesco: Zoographic Ambivalences in Paolo Mantegazza, Ouida, and Vernon Lee." *Gothic Studies* 16, no. 1 (2014): 111–127.

Robinson, Victor. "Introduction". In *The Sexual Relations of Mankind*, edited by Paolo Mantegazza and translated by Samuel Putnam. New York: Eugenics Publishing Co., 1935.

Secord, James. "Knowledge in Transit." *Isis: A Journal of the History of Science* 95 (2004): 654–672.

Sigusch, Volkmar. "The Birth of Sexual Medicine: Paolo Mantegazza as Pioneer of Sexual Medicine in the 19th Century." *Journal of Sexual Medicine* 5 (2008): 217–222.

Starr, Frederick. "Sketch of Paolo Mantegazza." *Popular Science Monthly* 43 (1893): 549–551.

Topham, J. R. "Introduction Focus: Historicizing 'Popular Science'." *Isis: A Journal of the History of Science* 100 (2009): 310–318.

Turbil, C. "Paolo Mantegazza and the Dream of 'Making' Science Popular." *Public Understanding of Science* (2017): 1–5. http://journals.sagepub.com/doi/full/10.1177/0963662517695117. Accessed 29 June 2018.

CHAPTER 11

The State and the Stomach: Feeding the Social Organism in 1830s New England

Molly S. Laas

Could abstaining from meat cause social unrest? The question became a pressing one for American medical practitioners in the 1830s, as they knew that what one consumed and digested had consequences. The medical literature of the late eighteenth and early nineteenth centuries held that a problem arising from indulging in the wrong kind of foods could transmit disturbances to the rest of the body, causing both illness and, potentially, immoral behavior. Foods were understood to stimulate or soothe the body, so meat, alcohol, or spices eaten in excess might provoke nervous excitement. This was a fundamental point for vegetarian health reformers in the period, who argued that stimulant foods were so dangerously exciting that they could sicken the body. New Englanders' meat-eating habits rendered them just one beefsteak away from total nervous exhaustion. Orthodox physicians, by contrast, held that most people needed the gentle stimulus of meat, particularly when they were tired or overworked, and that the regional custom of eating meat at all three meals per day ensured health and strength. While the question of

M. S. Laas (✉)
University of Göttingen Medical School, Göttingen, Germany

M. Mathias and A. M. Moore (eds.), *Gut Feeling and Digestive Health in Nineteenth-Century Literature, History and Culture*, Palgrave Studies in Literature, Science and Medicine,
https://doi.org/10.1007/978-3-030-01857-3_11

meat-eating bore intimately on the health of individuals, medical practitioners understood that the choice of diet had social ramifications, hinging on whether 1830s New England was inherently pathological. Did good health spring from keeping one's distance from a sick society and tending to one's own internal economy with a vegetable diet, as health reformers alleged? Or, was well-being a social construction, as orthodox physicians held, and health came from integrating oneself more fully into the customs and habits of one's society?

This paper will explore the question by examining a debate in 1835 about how to feed the social organism between the health reformer Sylvester Graham and the physician Luther Bell. The debate was instigated by Graham's speeches to the working class of New England in the early 1830s, in which he sought to convince workers that abstaining from meat and other stimulant foods was the only way to preserve health. Bell's essay was written as a corrective to Graham's vegetarian provocations, with the intent of reacquainting New England workers with the proper regimen for their station in life. It had been chosen by a prize committee sponsored by Harvard University and the *Boston Medical and Surgical Journal* as the best response to the "alimentary radicals" stirring up New England workers.[1] For Bell and for many of his fellow physicians, Graham's ideas had the potential to inflict wider social shifts than just a smattering of people sustaining themselves on brown bread and fruit. The importance of the region's diet went beyond individuals' choice of regimen: The people who made up New England had to be healthy so that the society of New England retained its robustness. As abstaining from meat was so contrary to received notions of personal self-management, some physicians argued that vegetarianism posed a destabilizing threat to the social system. It was entirely possible that health reformers could induce enough people to follow an erroneous diet that could weaken them, therefore damaging the well-being of society as a whole.

Health reform in the antebellum period is often portrayed in medical history as open warfare between the organized medical profession and alternative practitioners, as they fought for patients in the medical marketplace.[2] What can be effaced in this portrayal is the fact that both

[1] "Editorials and Medical Intelligence."

[2] Paul Starr, *The Social Transformation of American Medicine*, 52; Adam D. Shprintzen, *The Vegetarian Crusade: The Rise of an American Reform Movement, 1817–1921*; and James C. Whorton, *Nature Cures: The History of Alternative Medicine in America.*

alternative practitioners and orthodox physicians drew from some of the same basic medical theories and assumptions, particularly about the direct relationship between food and morality. On a deeper level, their arguments reveal their differences to be fundamentally about how each group understood the health of the social organism. First fully articulated by Comte in the 1830s, the social organism concept had roots in both social theory and in biological theory. It also had wide currency as an analogy useful for thinking about the relationship of individuals to society. Questions of national diet brought the analogy to the fore. For example, nineteenth-century British discussions of dyspepsia treated it as a national disease, where the stomach of the individual suffering Briton stood in for a social malaise or a maladjustment of the people who lived (queasily) within British society.[3]

In the USA, the question of feeding the social organism depended on how healthy one thought the organism was in an age of industrialization and urbanization. A perspective grounded in American exceptionalism portrayed Americans as uniquely strong and healthy. Some Americans believed that their country was a singular place, peopled by landowning farmers and prosperous artisans and blessed with so much fertile land that the ample food supply would always outstrip population growth.[4] These putative aspects of the country meant that it would be spared the class conflict and poverty that had attended urbanization and industrialization in Europe.[5] This widespread conviction was accompanied by anxieties that the age of American blessedness was beginning to pass away. Americans grew richer, cities grew larger, and salutary agrarian labor was becoming less common. Yet people working indoors were still eating like they were farm workers: too much, too quickly, and with gallons of liquor to wash it all down.[6] It was possible that this diet needed radical change, a return to a more natural form of living to counteract social corruption. The crux of the debate between Graham and Bell, then, was whether society was sick, and vegetarianism was the cure, or if society was healthy, and further adherence to its strictures would ensure personal well-being.

[3] Ian Miller, *A Modern History of the Stomach*, 19.

[4] Steven Stoll, *Larding the Lean Earth*, 34.

[5] Dorothy Ross, *The Origins of American Social Science*, 22–28.

[6] James C. Whorton, *Crusaders for Fitness*, 27.

Sylvester Graham, trained as a Presbyterian minister, found his calling as an itinerant temperance lecturer and self-taught expert in the science of human life. Though Graham did not acknowledge any intellectual or medical predecessors, his ideas about proper dietary regimen drew upon vitalist medical theories that considered the stomach to be the most essential organ for assuring health.[7] These theories were articulated by a range of physicians in the eighteenth century: Samuel August Tissot, François Broussais, John Abernethy, William Cullen, and Cullen's American student Benjamin Rush.[8] According to Graham's synthesis of these ideas, the central guarantor of a person's health was the proper functioning of the nerves at the pit of the stomach, which was the "great, primary, and common center of organic life."[9] These nerves directed the system's requirements for air and food and acted without recourse to the conscious will of a person, just as all organs in a healthy person functioned without thought. The nerves of the stomach further had relations with other, smaller clusters of nerves in the other organs. If the stomach was disturbed, say, by eating "substances which are not adapted to the wants of the vital economy," like meat or alcohol, waves of irritation would radiate out to the other organs of the body, leading to disease in those areas.[10] Further, if disturbances of the mind—such as nervousness or mental exertion—upset the functioning of the stomach, a general morbid irritability could render the nervous system extremely excitable and prone to illness.

For Graham, such was the state of New England man in 1830s America: His simple, natural desires had been distorted by the artificiality of society, which led him to indulge in dangerous stimulants like meat, spices, and alcohol. Even worse, he ingested twice the amount of food required by the animal economy. The laws of health called for sustaining oneself on only vegetables and grains, the least stimulating foods. They also mandated bathing in cold water and abstaining from sexual activity. Avoiding these stimuli would calm the nerves at the pit of the stomach and diminish the stresses of life.[11] Graham's regimen was a

[7] Stephen Nissenbaum, *Sex, Diet, and Debility in Jacksonian America*.

[8] Miller, *A Modern History of the Stomach*, 12–13; Nissenbaum, *Sex, Diet, and Debility in Jacksonian America*.

[9] Sylvester Graham, *Lectures on the Science of Human Life*, 180–181.

[10] Graham, 1, 350–352.

[11] Ibid., 195.

response to the particular anxieties of the Jacksonian era, as it provided a remedy for people who had fallen ill due to their reliance on a corrupt society and the capitalist economy. His dietetic prescriptions were therefore the route for a retreat from society. The harried businessman of the 1830s could find relief from his worries by pulling back from active life, washing in cold water, sitting down, and slowly chewing a hard biscuit.[12]

Graham's best living exemplar of the untainted appetite was the "wild boy" Kaspar Hauser. In 1828, the 17-year-old Hauser had appeared in Nuremberg. The boy said he had been imprisoned in a cellar from as far back as he could remember, and while he had received bread and water from an unknown, unseen caretaker, he had no human contact and so could only speak a few words. Hauser's appearance was an international sensation, and Graham's portrayal of Hauser as the epitome of man in a state of nature went along with the popular literature on the boy, which portrayed him as an innocent victim of the powerful and an avatar of naturalness.[13] Hauser was particularly noted for having an acute sense of smell and taste. New foods were revolting to him, particularly any kind of meat, milk, alcohol, coffee, or heavily spiced dishes. Owing to the concurrence of Hauser's preferences with Graham's dietetic strictures, Graham saw his sensory acuteness as resulting from his untainted, natural appetite, reared as he was on bread and water and kept from the debilitating effects of society. "Caspar's extraordinary acuteness and power of hearing were in no degree the effect of cultivation, but depended entirely on the pure, natural sensibilities of his organs: or on the very great degree of natural and healthy sensory power of his nervous system," Graham wrote.[14] As Hauser became accustomed to life outside the basement, the acuteness of his senses diminished, which Graham blamed on the enervating effects of meat consumption and exposure to unnatural and overstimulating society. In Graham's view, health sprung from cultivating a separateness from the world and a calmness of affect that verged on inertia.[15] Health, therefore, would come from retreat. If it was impossible for the excitable person of the 1830s to achieve the

[12]Nissenbaum, *Sex, Diet, and Debility in Jacksonian America*, 28.

[13]Martin Kitchen, *Kaspar Hauser: Europe's Child*, 179.

[14]Graham, *Lectures on the Science of Human Life*, vol. 2, 298.

[15]Nissenbaum, *Sex, Diet, and Debility in Jacksonian America*, 133–134.

naturalness embodied by Kaspar Hauser, he could at least reject meat and society alike in order to regain some of what had been lost.

Central to Graham's regimen was a theory of what foods were natural to man. Graham derived his theory from his structural view of the human body and the bodies of non-human primates, which he believed contained hints about the natural dietetic character of man. For Graham, non-human primates were the closest that people of the 1830s could get to understanding original man in a state of nature. He compared the morphologies of the skull, jaw, and teeth of humans to those of vegetarian non-human primates to argue for the naturalness of a diet devoid of meat and other stimulants.[16] Graham dismissed efforts by comparative anatomists to compare the corner teeth of the human jaw with the sharp cuspids of carnivorous animals, which were fit only to tear the flesh of prey and swallow it whole. He argued that since this was not a human practice, it made more sense to compare human teeth with the grinding molars of herbivores. Notably, the teeth and jaws of orangutans were the closest to those of man. In a "perfectly pure state of nature, when left free to choose their own nourishment and follow their undepraved instincts," these animals ate only fruit, Graham enthused.[17] Orangutans could be induced to eat flesh in captivity, however, leading some observers to argue that this fact meant they were omnivorous. Graham contended this argument was a "monstrous absurdity," illustrating the corruption by society of natural vegetarian instincts.[18]

In his evocation of the "state of nature," Graham was in line with the romanticism of his fellow American health reformers, who portrayed their advice as constituting a more natural way of living. Once a natural regimen had been achieved, man would be purified and returned to a state of nature.[19] This romanticism coincided with romantic strains in the physiological literature that the health reformers had adopted as foundational to their theories. The Swiss medical theorist of irritation Samuel August Tissot, for example, who was a friend and correspondent of Jean-Jacques Rousseau, shared with the philosopher a vision of contemporary

[16] It is possible that Graham had an opportunity to see an orangutan himself or read about it in the local periodicals, as a captive one was in Philadelphia in 1836. "Description of an Hermaphrodite Orang-Outang Lately Living in Philadelphia."

[17] Sylvester Graham, *Lectures on the Science of Human Life*, vol. 2, 68.

[18] Ibid.

[19] Whorton, *Crusaders for Fitness*, 30.

life as pathological, full of excessive stimulation which engendered a dangerous nervous sensibility, and he argued that a return to natural health would cure social problems.[20] For Rousseau, to understand people's true nature, one had to reconstruct original man; all that we can know about human nature must come from separating out what was artificial from what was natural, with the natural being the most purely and correctly human.[21]

Graham was likely to have encountered Rousseauian ideas about nature through his reading of Tissot, and his evocation of a "state of nature" bore striking similarities to that of the French philosopher. Rousseau's designation of vegetarian food as the natural diet of man stemmed from his belief that man in his natural state would be innocent of violence, and therefore would have eschewed meat.[22] The fact that even uncivilized men killed for food was proof for Rousseau that people everywhere had irrevocably left the state of nature. Likewise, Graham thought that all human beings were unnatural: The foods they ate were "so artificial that there is nothing about them that could be considered as part of the natural dietetic character of man."[23] As a result, he suggested, there were no true men left: All societies were so corrupt that "so far as [nutrition] is anatomically considered, man must in strict propriety be regarded as an extinct species."[24] If original man had died out, there was at least a process of repair for his enervated descendants. Unlike Rousseau, who despaired of any attempt at a return to the state of nature, Graham offered his own system as a means for contemporary people to create a contemporary simulacrum of the original state of nature. Brown bread, cold water, and fresh fruit were the foods of innocence. Recovering that innocence was the only means for obtaining pure and natural health.

Graham's vision of the modern body, therefore, was one that had degenerated from the pinnacle of health, but could be restored if a placid mind, nerves, and stomach were maintained by gentle, solitary labor and a calming diet. It held that God had designed human beings with fixed

[20] Anne Vila, *Enlightenment and Pathology*, 187–188.

[21] Francis Moran, "Between Primates and Primitives: Natural Man as the Missing Link in Rousseau's Second Discourse."

[22] Jean-Jacques Rousseau, *The Discourses and Other Early Political Writings*, 193–194.

[23] Graham, *Lectures on the Science of Human Life*, vol. 2, 51.

[24] Ibid.

relations among the physical, moral, and mental faculties. This arrangement would remain in harmony if all of the laws of health were obeyed, and among these laws was the imperative to protect oneself from external dangers. Graham employed a state-body analogy to draw attention to the relationship between the individual parts of a body and the larger whole, an analogy that underscored the risk of external threats. Errors in self-protection—such as exposure to the debilitating effects of contemporary life or external, disease-causing agents—were like the foreign enemies that could assail a state, Graham wrote. In Graham's view of the social organism, the body's organs were akin to the "husbandman, and artisan and merchant, and other members of the commonwealth," who would be roused to fight for the good of their nation during an invasion.[25] More than an articulation of Graham's medical ideas, this metaphor expressed his populism. Notably, the model he described lacked any sort of governing figure that might direct the action of the organs/citizens; not even the stomach, which was for Graham the most important organ for maintaining health, could act as a leader for the rest of the body's assault on disease. Instead, each organ acted independently for the good of the whole; the metaphor expressed Graham's faith in the republican virtue and sound instincts of the common man in the absence of government. Further, Graham's vision of the body as a state countenanced no alliances, nor the possibility of a confederacy of nations. Each person must tend to his own bodily economy alone, without reference to the society in which he lived. If society was artificial, and therefore only a source of corruption, natural health could only be attained by maintaining distance from one's fellows.

The physician Luther Bell took a very different view of human health, a view that stemmed from his medical training and his philosophical standpoint. Bell is best known to historians of American medicine for his years as the medical superintendent of the McLean Asylum for the insane in Massachusetts, but prior to obtaining that post in 1837 he was a country doctor and legislator in New Hampshire. Bell situated himself at the vanguard of American medicine in a time of great change. In the 1830s, some physicians were beginning to question prevailing medical theories. The therapies that these theories indicated, like extensive bloodletting and the administering of dangerous drugs like calomel,

[25] Ibid., 137–138.

were not only useless in curing disease but were actively harmful.[26] These doctors were influenced in part by the theoretical framework of the Paris clinical school, particularly the commitment to understanding disease through pathological anatomy, and its skepticism about existing theories and treatments. American physicians aligned with the Paris Clinic held that until more accurate facts were known about the true nature of disease, physicians were left with few therapies to draw from and generally preferred a noninterventionist approach, relying on good food and rest to support the body's ability to heal itself.[27] A notable percentage of elite American physicians traveled to Paris to study medicine in the 1830s, and while Bell was not among them he shared their skepticism of medical interventions. In an 1836 essay he wrote on physiology's relationship to the understanding of disease, Bell roundly dismissed the medical theorists of the past and their purported knowledge of how to heal. In particular, Bell reserved his harshest scorn for the theorists that Graham relied on most heavily in his discussion of irritation—Cullen, Rush, and Broussais—noting that their physiological facts were false, their theories unprovable, and their therapies actively harmful to health. Further, he argued that adherents to their theories had been led away from the discoveries of pathological anatomy, a false path which had resulted in "incalculable injury to man, and to the progress of science."[28] Instead, doctors like Bell proposed limiting the drugs and procedures they prescribed to their patients, instead seeking to support the body's natural ability to heal itself. In his later work as the head of the McLean Asylum, Bell decreased the number of medical interventions patients underwent and instead furnished decent food and recreational and occupational therapy in the hopes that gentle treatment would be a more effective cure than restraints and drugs.[29]

[26]Bloodletting and the mercury compound calomel were used to purge the body and "shock" it out of a humoral imbalance. Repeated bloodletting, and the administering of large doses of purgative drugs, was known as "heroic medicine" and was a subject of criticism by both orthodox physicians and alternative healers alike. See John Duffy, *From Humors to Medical Science*, 72.

[27]John Harley Warner, *The Therapeutic Perspective: Medical Practice, Knowledge, and Identity in America, 1820–1885*, 22.

[28]Oliver Wendell Holmes, Luther Bell, and Robert W. Haxall, *Dissertations on the Question*.

[29]S. B. Sutton, *Crossroads in Psychiatry*, 58.

Bell's challenge to medicine led him also to reject Graham's notion that gastric irritation was the central root of bad health. His own model of health described the human organism as situated within a series of nested relationships, culminating in the health of society as a whole. On the innermost level, an individual's intellectual and moral facilities were dependent upon the proper functioning of the body, and a body required a correct diet to support thought and moral action.[30] Unlike Graham, Bell saw an individual's maintenance of his health as having social ramifications. This was the outermost ring of the nested relationships, in which the individual's well-being formed part of the health of the social body overall. In linking social and individual health, Bell expanded on physicians' traditional definition of their profession. More broadly, he was writing at the beginning of the public health movement in Britain, which also sought a social definition of health; yet such developments were only nascent in the USA, and in neither Britain nor the USA was diet a central aim of early public health advocates. Bell was self-aware about his effort to expand the scope of his profession. In considering diet and nutrition, physicians had previously only considered the habits of the leisured classes, because they had been up to this point the main consumers of diet advice. Wealthy people had the luxury to choose their foods and tended to corpulence due to overindulgence, and so were in need of medical guidance. Bell noted that the shift toward studying the nutrition of laborers was an odd choice on the surface, as the working class were far less liable to committing errors in diet. They tended to be engaged in exercise and lacked the surplus money that would allow them to stuff themselves on luxuries, and so were healthfully constrained in their choices and liable to work off any excess food.

It was a shift in economic thought, not medical thought that caused doctors to consider the diet of the working class, Bell argued. He noted that political economists working on the question of population were beginning to think of man as a "laboring animal," which linked a people's health to their productive power.[31] Therefore the question of the public's diet was an essential one for understanding the economy of a country, which depended upon the physical capacity of its people. That Bell made this point is notable, in that it indicated his broader

[30] Luther V. Bell, *A Dissertation on the Boylston Prize-Question for 1835.*

[31] Bell, *A Dissertation on the Boylston Prize-Question for 1835*, 4.

acceptance of political economy as part of the worldview of medicine. This was not an obvious connection, as political economy and medicine were seen by some to work at cross-purposes. The object of political economy was to promote the general good of the community at the possible expense of individuals, and the object of medicine was to protect the health of the most feeble without regard to the welfare of the community. Thus, the two disciplines were in conflict.[32]

Yet Bell saw no such conflict. This was likely due to the broader intellectual and political culture of the USA that caused a doctor like Bell to consider the welfare of others to be within his sphere of professional interest. Bell had been educated at Bowdoin College in Maine. His education had been imbued with the moral philosophy taught in American colleges in the period, which knitted together metaphysics, religion, and the study of politics and political philosophy.[33] Stemming from eighteenth-century Scottish Common Sense philosophy, it posited that the external world was knowable and objective. God made common sense as a tool for people to use, and therefore men could trust what they saw and could understand nature.[34] This philosophy was designed to be the capstone of the American college curriculum, inculcating students with republican virtues and trust in a person's rational faculties.[35] Politically, too, Bell was aligned with the American Whig party, which instantiated common-sense moral philosophy by casting itself as the party of national development, central government, restraint, and self-control.[36]

Bell's education had provided him with a confidence in man's capacity for reason, and his political party stressed social and political intervention in the service of progress. A third influence on his thinking about diet was the temperance movement. The organized temperance movement

[32] John Ayrton Paris, *A Treatise on Diet*, 127; Christopher Hamlin and Kathleen Gallagher-Kamper, "Malthus and the Doctors: Political Economy, Medicine, and the State in England, Ireland, and Scotland, 1800–1840"; and Christopher Hamlin, *Public Health and Social Justice in the Age of Chadwick: Britain, 1800–1854*, 25.

[33] George Edward Ellis, *Memoir of Luther V.* His classmates at Bowdoin included the novelist Nathaniel Hawthorne, the poet Henry Wadsworth Longfellow, and the US president Franklin Pierce.

[34] Gerald N. Grob, *Edward Jarvis and the Medical World of Nineteenth-Century America*, 15; Ross, *The Origins of American Social Science*, 36.

[35] Henry F. May, *The Enlightenment in America*, 346–348.

[36] Isaac Ray, *A Discourse on the Life and Character of Dr. Luther V. Bell*; Daniel Walker Howe, *The Political Culture of the American Whigs*, 33–35.

had just begun in 1826 with the founding of the American Temperance Society.[37] Bell was no stranger to the cause; he had been the president of a New Hampshire temperance organization. In that post, he advocated total abstinence from all types of alcohol, and so he had deep sympathies with temperance reformers' efforts to make the dangers of alcohol known to the broader public.[38] In his essay on diet, Bell noted that the temperance movement had brought about an "extraordinary and truly unparalleled change" in the daily habits of Americans.[39] He credited them with alerting Americans to the error of drinking alcohol, the unhealthfulness of which they were heretofore completely unaware; it also caused them to think about the dangers that might lurk in the foods they commonly consumed. In essence, the temperance movement helped effect a shift in medical concerns about diet in the USA from quantity to quality. As the historian Elizabeth Williams has noted, Cullen thought that the appetite could become diseased or disordered, with the result of causing over or under-eating.[40] By contrast, American diet reformers of the 1830s and 1840s were far more concerned with what was eaten rather than how much, as they had translated the rhetoric and fervor of the temperance movement to diet. Alcohol, meat, and rich pastry were considered to be more threatening to health than the amount of food consumed.

In his essay on diet, however, Bell attempted to push back against the notion that meat was just as bad for health as liquor by insisting that meat was necessary for New England workers in the climate they lived in. It was a commonplace of medical science of the period that the body was in a relationship with its environment, and it was thought that environmental problems (like smelly effluvia or marshland) might harbor dangers to the body.[41] In the case of diet, there was a rule of thumb relating climate to the types of food people ate. For Bell, the natural and social environment was the main indicator for a healthy diet. Even the

[37] Matthew Warner Osborn, *Rum Maniacs: Alcoholic Insanity in the Early American Republic*; Ian R. Tyrrell, *Sobering Up: From Temperance to Prohibition in Antebellum America*.

[38] Bell, "An Address Before the Young Men's Entire Abstinence Association," 7.

[39] Bell, *A Dissertation on the Boylston Prize-Question for 1835*, 6.

[40] Elizabeth A. Williams, "Sciences of Appetite in the Enlightenment, 1750–1800."

[41] Conevery Bolton Valenčius, *The Health of the Country: How American Settlers Understood Themselves and Their Land*, 159–160.

Bible, he noted, contained a clear and comprehensive dietary code for the ancient Israelites to follow, which had been "adapted with wonderful sagacity and wisdom to the climate and to the habits of the Israelites," preserving their health throughout their years in the desert. Bell did not argue that since this diet had been ordained by God it was therefore useful for nineteenth-century New Englanders as well as ancient Israelites; instead, it had to be fitted to the circumstances of the moment. A healthy diet for New Englanders ought to suit their local conditions. As far as Bell was concerned, this was simple: New England rested in the temperate belt of the globe, where people ate a mixed diet of meat and vegetables. The moderate stimulus of animal flesh helped the industrious New England laborer survive in a cold climate, but as the weather was not so cold as regions farther north, only meat once a day was needed to support heat and work. This meant that the diet of New England should be exactly what it was: meat, bread, and vegetables, though not too much of any one category. Following in the footsteps of centuries of physicians before him, Bell recommended adherence to the Galenic non-naturals, notions of moderation and restraint that were the true guides to both good diet and good health.

Bell's emphasis on culture and climate in his diet prescription meant that the question of what food was "natural" in itself had no meaning. Of all the claims that Graham made in his lectures, the one that Bell found most provoking was the notion that man had lost his essential character over the course of years of social development. For Bell, health itself was a product of this development. Graham's contention that a feral childlike Kaspar Hauser should exemplify the height of salutary living was nothing but a provocation to Bell. For him, a strong connection to one's society and a continual striving for betterment were the cornerstones of a life well lived. A person who was without any sort of social instruction was practically irredeemable. Bell devoted a notable part of his essay on diet to describing and analyzing the life of "Wild Bill, or the Mississippi Orson" a white boy who had been found naked and alone in the Mississippi wilderness.[42] Bell reprinted an article from the *Knickerbocker* about this boy, who had lived alone in a Mississippi

[42]The name "Orson" in the title of the article referred to the medieval romance, *Valentine and Orson*, about twin brothers abandoned in the wilderness in infancy. Valentine is found and raised as a knight; he eventually locates his brother, Orson, a wild man in the woods, and tames him.

swamp until he was found by settlers in 1811. When he was discovered, the child was about nine years old, and had survived by hunting frogs and small animals and devouring them raw. He ran from the men who first encountered him and was finally captured after being hunted down by dogs. Unlike Hauser, who had the benefit of a careful education once he was found, Wild Bill had been thrown among a rough frontier people who were unwilling to teach him the nuances of social life or to place him in school. "In a forest full of Indians and wild animals, Wild Bill was an object of very little higher interest than a tamed bear or panther," for his neighbors, the article noted. As he was poorly educated and hardly civilized, Wild Bill exemplified the qualities of his name even after he had lived among people for some years. He could barely talk, had a fierce and violent temper, and hated to wear clothing and to work. He retained his preference for uncooked meat, and developed a passion for liquor, especially when it was sweetened. Thus, his appetite remained savage and untutored, with a strong preference for stimulating foods like meat and alcohol; a far cry from the notion that man was naturally vegetarian.[43]

Bell argued that Wild Bill more closely approximated the "homo naturæ" than Kaspar Hauser did, as the German child had been clothed, fed, and kept indoors during his period of isolation and therefore had some human care. The fact that Hauser was repulsed by meat was simply a result of being raised on a vegetable diet, not a sign of the instinctive inclinations of the innocent palate. Bell argued that "natural tastes" did not exist; the conventional customs of any society shaped preferences to the point where there would be no distinguishing between a person's natural appetite and his acquired taste. In his response to Graham, Bell attacked Rousseau head on; he equated Grahamism with the French philosopher's notion that man living in a state of nature outside of society was fundamentally innocent of corruption. In Bell's view, this was a false understanding of nature. A person left alone in the world would be more like Wild Bill: savage, brutish, and incapable of discrimination. Raw, unrefined qualities in the natural world and in human beings must be purified through reason, education, science, and commerce in order to be good. Likewise, man must maintain his moral and physiological balance by making all of his habits compatible with both the society he lived in and the best advice from physicians of authority.

[43] Bell, *A Dissertation on the Boylston Prize-Question for 1835*, 20–23.

In keeping with his broader skepticism of elaborate medical or scientific theory, Bell further rejected Graham's notion that non-human primates exemplified some sort of missing link between contemporary human beings and the rest of the animal kingdom. Against Graham's evocation of the close kinship between man and apes, Bell argued that humans, no matter if they were in a civilized or uncivilized state, constituted "a species *sui generis*, far removed from all others."[44] Therefore, Graham's contention that the teeth and jaw of non-human primates offered the best guide to the diet of man was completely irrelevant to Bell. While he agreed with the general view of comparative anatomists that the teeth, jaws, and alimentary canals of animals were suited to the kind of food they ate, Bell contended that such details were insufficient for making any solid conclusions about the natural diet of man. Instead, Bell argued, the analogy of form of humans' teeth, jaws, and digestive tract to that of apes showed only that these creatures could eat food with a similar consistency to soft, ripe fruit. All human societies ate both animal and vegetable food in a prepared state; cooked or otherwise refined to make it soft enough to chew and digest. This showed, Bell wrote, that man was designed "to be governed in his food as in everything else, not by instinct, but by his reasoning faculties," which allowed him to "subdue nature to himself."[45] Therefore, there was no need to assume that only sharp-toothed carnivores could eat meat; man could chew cooked meat with his herbivore jaw.

In arguing that what is "natural" had no bearing on human culture and choices, Bell drew on the lectures on physiology and the natural history of man by the English surgeon William Lawrence. Lawrence noted the differences between man and animals, particularly humans' long infancy and the fact that all people across the globe live in societies. Both traits lifted human nature above that of animals. Social life and progressive civilization, Lawrence wrote, were "very valuable parts of his nature, as much as the erect stature and speech."[46] Humans would not be human if they lacked social intercourse, civilization, and language, and so it made no sense to assume that there had existed men in a state of nature, let alone to posit that a vegetarian diet was the proper food of

[44] Bell, *A Dissertation on the Boylston Prize-Question for 1835*, 12.

[45] Ibid., 7–12.

[46] William Lawrence, *Lectures on Physiology, Zoology, and the Natural History of Man, Delivered at the Royal College of Surgeons*, 187.

natural man.[47] For Bell, the central point in designing a diet was to have recourse to human reason, within the context of the human society one lived in. New Englanders should maintain their health by aligning their habits with that of their neighbors, not of feral children or wild apes.

Both Bell and Graham saw people's diets as a central part of the healthy functioning of the social organism, but they differed on the structure and function of the organism's parts. Graham's view of the social organism emphasized independence to the point of isolation; his thinking can be seen to be in line with a robust strain of libertarianism in American political and social life. His vision of health rested on the practice of eating unstimulating foods and pulling away from active social life, tending to one's own bodily needs without recourse to others and attempting to recreate man's pre-social state of nature. Bell's social organism, by contrast, emphasized interconnectedness and the power of social interaction to educate and refine its members, reflecting his political Whiggery and its emphasis on centralized governance. The underlying vision of society for Bell was analogous to that of a body in perfect balance, with his passions under proper control and reason directing the action of the system. A person who did not regulate his diet and emotions was just the same as a country in the grip of *laissez-faire* social policy, which left unmanned the vital responsibilities of the state. So it follows that a mass of New Englanders, riled up by Graham's vegetarian provocations, were seen in Bell's eyes to pose a threat to the harmoniousness of New England society. In offering diet advice to New England workers, Bell sought to convince his audience to bring their own bodies into harmony with the society in which they lived, in order to ensure both their own health and the health of New England as a whole.

Bibliography

Bell, Luther V. *A Dissertation on the Boylston Prize-Question for 1835: What Diet Can Be Selected Which Will Ensure the Greatest Probable Health and Strength to the Laborer in the Climate of New England?—Quantity and Quality, and the Time and Manner of Taking It, to Be Considered.* Boston: D. Clapp, Jr., 1836. https://collections.nlm.nih.gov/catalog/nlm:nlmuid-101163642-bk.

[47] William Lawrence, *Lectures on Comparative Anatomy, Physiology, Zoology, and the Natural History of Man*, 184, 197.

Bolton Valenčius, Conevery. *The Health of the Country: How American Settlers Understood Themselves and Their Land*. New York: Basic Books, 2002.

Duffy, John. *From Humors to Medical Science: A History of American Medicine*. 2nd ed. Urbana: University of Illinois Press, 1993.

"Editorials and Medical Intelligence." *The Boston Medical and Surgical Journal* 13, no. 14 (11 November 1835): 223–228. https://doi.org/10.1056/NEJM183511110131407.

Ellis, George Edward. *Memoir of Luther V. Bell, M.D., LL. D.: Prepared by Vote of the Massachusetts Historical Society*. Boston: J. Wilson and Son, 1863.

Graham, Sylvester. *Lectures on the Science of Human Life*. Vol. 1. Boston: Marsh, Capen, Lyon and Webb, 1839. https://collections.nlm.nih.gov/catalog/nlm:nlmuid-61411880R-mvset.

Grob, Gerald N. *Edward Jarvis and the Medical World of Nineteenth-Century America*. Knoxville: University of Tennessee Press, 1978.

Hamlin, Christopher. *Public Health and Social Justice in the Age of Chadwick: Britain, 1800–1854*. Cambridge: Cambridge University Press, 1998.

Hamlin, Christopher, and Kathleen Gallagher-Kamper. "Malthus and the Doctors: Political Economy, Medicine, and the State in England, Ireland, and Scotland, 1800–1840." In *Malthus, Medicine, & Morality: "Malthusianism" After 1798*, edited by Brian Dolan. Wellcome Institute Series in the History of Medicine. Amsterdam: Rodopi, 2000.

Holmes, Oliver Wendell, Luther Bell, and Robert W. Haxall. *Dissertations on the Question: How Far Are the External Means of Exploring the Condition of the Internal Organs to Be Considered Useful and Important in Medical Practice? For Which Premiums Were Adjudged by the Boylston Medical Committee of Harvard University*. Boston: Perkins & Marvin, 1836. https://archive.org/details/63640600R.nlm.nih.gov.

Howe, Daniel Walker. *The Political Culture of the American Whigs*. Chicago and London: The University of Chicago Press, 1979.

Lawrence, William. *Lectures on Physiology, Zoology, and the Natural History of Man, Delivered at the Royal College of Surgeons*. Salem: Foote and Brown, 1828.

———. *Lectures on Comparative Anatomy, Physiology, Zoology, and the Natural History of Man*. London: Henry G. Bohn, 1848.

May, Henry F. *The Enlightenment in America*. New York: Oxford University Press, 1976.

Miller, Ian. *A Modern History of the Stomach: Gastric Illness, Medicine, and British Society, 1800–1950*. Vol. 4. Studies for the Society for the Social History of Medicine. London: Pickering & Chatto, 2011.

Nissenbaum, Stephen. *Sex, Diet, and Debility in Jacksonian America: Sylvester Graham and Health Reform*. Westport, CT: Greenwood Press, 1980.

Osborn, Matthew Warner. *Rum Maniacs: Alcoholic Insanity in the Early American Republic*. Chicago and London: University of Chicago Press, 2014.

Paris, John Ayrton. *A Treatise on Diet*. London: Printed for Thomas and George Underwood, 1826.

Ray, Isaac. *A Discourse on the Life and Character of Dr. Luther V. Bell: Read to the Association of Superintendents of North American Institutions for the Insane, at Its Annual Meeting, in Providence, R.I., June 10th, 1862*. Boston: J.H. Eastburn's Press, 1863. http://archive.org/details/9609410.nlm.nih.gov.

Ross, Dorothy. *The Origins of American Social Science*. Ideas in Context No. 19. Cambridge: Cambridge University Press, 1990.

Shprintzen, Adam D. *The Vegetarian Crusade: The Rise of an American Reform Movement, 1817–1921*. Chapel Hill: The University of North Carolina Press, 2013.

Starr, Paul. *The Social Transformation of American Medicine*. New York: Basic Books, 1982.

Stoll, Steven. *Larding the Lean Earth: Soil and Society in Nineteenth-Century America*. New York: Hill and Wang, 2002.

Sutton, S. B. *Crossroads in Psychiatry: A History of the McLean Hospital*. Washington, DC: American Psychiatric Press, 1986.

Tyrrell, Ian R. *Sobering Up: From Temperance to Prohibition in Antebellum America, 1800–1860*. Vol. 82. Contributions in American History. Westport, CT: Greenwood Press, 1979.

Vila, Anne. *Enlightenment and Pathology: Sensibility in the Literature and Medicine of Eighteenth-Century France*. Baltimore: Johns Hopkins University Press, 1998.

Warner, John Harley. *The Therapeutic Perspective: Medical Practice, Knowledge, and Identity in America, 1820–1885*. Cambridge: Harvard University Press, 1986.

Whorton, James C. *Crusaders for Fitness: The History of American Health Reformers*. Princeton, NJ: Princeton University Press, 1982.

———. *Nature Cures: The History of Alternative Medicine in America*. New York: Oxford University Press, 2002.

Williams, Elizabeth A. "Sciences of Appetite in the Enlightenment, 1750–1800." *Studies in History and Philosophy of Science Part C: Studies in History and Philosophy of Biological and Biomedical Sciences* 43, no. 2 (June 2012): 392–404. https://doi.org/10.1016/j.shpsc.2011.10.031.

CHAPTER 12

Food Faiths: Gut Science and Spiritual Eating

Catherine L. Newell

The fraught history of nutrition and gut health in the USA leads inevitably to one pivotal figure in the history of diet and wellness movements: Sylvester Graham. Although he is primarily remembered today for his eponymous cracker, his fusion of dietary philosophy with a distinctively American Presbyterian worldview shaped the course of both religion and diet from the Jacksonian Era of the 1830s until the end of the nineteenth century. Graham's dietary beliefs in the efficacy of brown bread, vegetables, and a total abstinence from meat and alcohol for human health and spiritual well-being consolidated a nascent religio-vegetarian movement by joining medical principles with existing Biblical-based vegetarian beliefs. Thanks to Graham, thinking of meat, alcohol, and caffeine as deleterious became part of the public mainstream. Tracing an historical line on the topics of food and health backward through the nineteenth century from almost any point inevitably leads to Graham's regime. For example, the great American medical innovator and diet reformer John Harvey Kellogg, whose cereals were

C. L. Newell (✉)
University of Miami, Coral Gables, FL, USA

M. Mathias and A. M. Moore (eds.), *Gut Feeling and Digestive Health in Nineteenth-Century Literature, History and Culture*, Palgrave Studies in Literature, Science and Medicine,
https://doi.org/10.1007/978-3-030-01857-3_12

one of the most remarkable food technologies at the beginning of the twentieth century, began his career as a devoted Seventh-day Adventist. One of the founders of Adventism, Ellen G. White, became a follower of Graham's principles after an illness and seeded the dietary philosophy into her theological reflections. These same principles were, for a time, put into practice at Oberlin College, a hotbed of American liberalism and social reform.[1] Vegetarian physician William Alcott, the first president of the American Vegetarian Society, also wrote extensively on the efficacy of the Graham System in various medical journals throughout the eighteen-thirties and forties; it was a dietary preference he shared with his second cousin and dear friend, Amos Bronson Alcott. In 1843, Bronson Alcott retreated to an experimental communal farm in rural Massachusetts, where he and his fellows attempted to create their own small Eden by living off the land and according to Grahamite principles.[2] This vegan commune was later eviscerated in the short story, "Transcendental Wild Oats," by Alcott's daughter Louisa May. The title's reference is to the Transcendentalism, the nineteenth-century American philosophical and spiritual movement that believed in the "transcendental" presence of the divine in nature and the human, which is held together with a single, pervasive force or "oversoul." Louisa May subtly derides these beliefs by suggesting in the story that the experiment failed because "about the time the grain was ready to house, some call of the Oversoul wafted all the men away."[3]

While nearly all of these moments occurred in the nineteenth century, they still have great resonance today. In particular, this essay explores their significance through an examination of Graham's science and the testimonials to his diet's efficacy. Most importantly, I will argue that Graham and his followers' recombination of physiological science in the service of spiritual well-being constitutes a form of religious belief that could easily be mistaken for scientific rationality. Finally, we will see how faith in Graham's dietary philosophy is illustrated in the first-person

[1]The experiment was short-lived; however, because while Oberlin's students and faculty were delighted to participate in acts of civil disobedience (among other perceived transgressions, Oberlin permitted women to study at the college and was a well-known stop on the Underground Railroad) a menu made up almost solely of brown bread and vegetables was a step too far.

[2]Francis, *Fruitlands*.

[3]Alcott, "Transcendental Wild Oats," 95.

narratives testifying to the effectiveness of his diet in helping his followers survive the American cholera epidemic of 1832.

Sylvester Graham: A Life

Graham might be the prototypical diet expert of today, save that he was born in 1794. His much-mythologized life began when he was born the seventeenth child of a Presbyterian minister, who died at the age of seventy-four (or seventy-two, depending on the source) two years after Sylvester's birth. Some of his biographers have found fodder for psychological analysis in the fact that, due to financial hardship and his mother's impending mental breakdown (she was institutionalized when Graham was eight or nine), not long after his father's passing, Graham was to live with a rotating cast of neighbors, relatives, and older siblings. Having been raised outside and away from the family hearth, as it were, Graham expended an extraordinary energy in later years on insisting that health began in the home with brown, unadulterated bread baked by "mother."[4] But before he became renowned for his lectures on bread and the homely hearth, Graham foundered for much of his youth, oscillating between bouts of illness and depression with efforts to hold down a respectable job and achieve an education. His elder brothers from his father's first family of ten children were well-respected, landed men with incomes and educations. The seven younger Grahams, on the other hand, were left to the winds of fate, and the youngest of them all suffered the most; it is no wonder that he struggled, or that he was, by his own admission, a rather difficult person.[5]

One significant element of his biography that is helpful in understanding his career choice as what today we would call a "lifestyle expert" is that his education was cut short by his own ill health and lack of a stable home. After an unhappy, deprived childhood and a series of menial jobs, by the time he was in his late twenties Graham was able to enroll in a college and begin studying to become a minister like his father. But just when he finally felt that he was headed toward a career (and, therefore, solvency), his fellow students at Amherst Academy—exhausted by his

[4] Nissenbaum, *Sex, Diet, and Debility*, 5 and ff. Much of the biographical material here (and in most studies of Sylvester Graham) comes from Nissenbaum's fundamental text on Graham's life and historical context.

[5] Nissenbaum, *Sex, Diet, and Debility*, 14.

posturing and self-importance—got him expelled on false charges, which precipitated a mental breakdown. But being forced to leave school also forced him to reconsider how he would make his way in the world. By the time he was in his early thirties, Graham had a wife, a family, and a revised calling: He decided to study on his own to become a minister and embark on a career as a traveling preacher. Once again, Graham had to rise above the circumstances that constrained him; in a nation fairly crawling with itinerant evangelists, Graham found his niche by accepting a position with the Philadelphia Temperance Society in 1830.

While not all ministers begin their careers with a church or an itinerant ministry, Graham's choice to work with the Temperance Society was appreciably different from the paths taken by other men of his era who trained to be ministers. If a church and a fixed congregation were unavailable, the next logical step was to establish oneself as an orator—something at which Graham excelled—who lectured on theological topics in a public space. But only those very devoted to a particular cause, such as temperance, passed up the opportunity to minister either within a church or to a larger community; volunteerism and social reform were often left to the laity, with the ministry providing inspiration and occasional leadership. Graham's affiliation with the temperance movement, however, did not necessarily indicate that he was passionate about preaching against the evils of alcohol; Graham himself confessed that he avoided alcohol primarily because he could not stand the taste. Graham was quietly sober for most of his life, despite living in an era when the majority of the country drank from morning through the night and consumed, among other spirits and fermented drinks, four or more gallons of rum per year.[6]

One plausible reason for Graham throwing in his lot with the Temperance Society, rather than carving out a niche elsewhere, was his interest in the medical sciences, particularly the physiological approach to health and wellness. Part of the appeal to Graham of the Temperance Society in Philadelphia was that it was organized not only by Quakers and Methodists, who oversaw similar societies in other cities and states, but that it was convened by medical doctors. Graham's layman's interest in physiology and medicine was put to use by incorporating recent

[6] Salinger, *Taverns and Drinking in Early America*, 2–4.

medical findings into his sermons for the Temperance Society. Soon this interest became more than fodder for his lectures, and in mid-1831 Graham left the society to devote himself full-time to studying and understanding the medical science of his day. Graham was not alone in his layman's interest in medicine. In addition to a particularly American mania for volunteer associations and improvement societies, Graham lived at a time when "democratic ideology received its sharpest expression in lay medicine."[7] Historian of American medicine Paul Starr, for example, counts Graham among the numbers of lay medical practitioners whose "informal practice" of medicine was "an extension of domestic care into the community." The societal and health reforms that shaped Jacksonian America were often powered by men like Graham, who studied medicine on their own reconnaissance and created a theory of human health that they then marketed to their public through speeches, pamphlets, and sermons (not unlike the diet culture of today with dietary advice marketed through books, blogs, and guest spots on television talk shows). Graham's self-taught medical knowledge and his gift for sermonizing soon converged on a subject that was peaking the interest of the American public: The importance of living a moral life not just for an eternal reward in the afterlife, but for the sake of achieving good health in this Earthly life—ease of digestion, life without physical pain, and the avoidance of diseases such as yellow fever, consumption (tuberculosis), and cholera.

A New Paradigm

Graham's interest in physiology occurred at a moment of a shift in the conception of the functionality of the human body. The paradigm of the four humors—yellow bile, black bile, phlegm, and blood—was slowly being augmented by the science of the nervous system. The relatively new science of neurology was introduced in American medicine by Dr. Benjamin Rush in the late eighteenth century.[8] It was Rush who helped inculcate the notion in eighteenth-century American medicine that the nervous system was the main source of debility, due to the presence of "sedative powers" reducing the brain's capacity to function,

[7] Starr, *The Social Transformation of American Medicine*, 47.

[8] Brodsky, *Benjamin Rush: Patriot and Physician*.

which in turn caused a breakdown of the body.[9] As with madness, so with all disease: Rush believed that the nerves extend to "the seat of the mind…[and when] preternaturally irritable, communicate more promptly, deranged action to the blood vessels of the brain."[10] Like most of his cohort, Rush was attempting to understand the mechanism of disease and realized that the humoral model routinely failed to account for a host of symptoms suffered by his patients. Rush also believed that a "state of debility could be caused by too much stimulation as well as by too little."[11] In other words, all symptoms—indeed all *disease*—could be traced back to the single handicap of an overstimulated nervous system.

The belief in Graham's era that the nervous system was holistically responsible for everything from indigestion to cancer to mental illness pervaded not only the medical sciences, but also the wider cultural landscape. Admirers of nineteenth-century literature, for example, undoubtedly recall the satiric trope of a character who is over attached to the notion that illness is caused by unhealthy nerves, as exemplified by Mrs. Bennet in *Pride and Prejudice* (1813) when she exclaims to her husband, "Oh, Mr. Bennet! How can you tease me so? Have you no compassion for my poor nerves?" (To which her exasperated husband replies, "Oh, you mistake me, my dear. I have the highest respect for them. They've been my constant companions these twenty years."[12]) In any case, Rush's scientific conclusions made him a respected advocate of temperance in all things, especially regarding alcohol, caffeine, and other stimulants. He advised instead the imbibing of warms drinks, "such vegetables as are least disposed to acidity," the avoidance of "ardent spirits," and the overconsumption of stimulants such as coffee and tea.[13] Disease, then, was the result of exterior forces on the nervous system, which was further evidence of the prevailing moral philosophy of the era—Protestant to its very core—that disease is the result of moral depravity, which in turn breeds an indifference to hygiene and a propensity to sin. Perhaps more than Rush's science of the nervous system, this idea—that disease is caused by immorality—is the hypothesis that most deeply influenced Graham's philosophy on food and drink.

[9] Brodsky, *Benjamin Rush: Patriot and Physician*, 91.

[10] Rush, *Medical Inquiries and Observations*, 101.

[11] Rush, quoted in Nissenbaum, *Sex, Diet, and Debility*, 55.

[12] Austen, *Pride and Prejudice*, 3.

[13] Rush, *Medical Inquiries and Observations*, 100–103.

Graham's studies in temperance and physiology started when he began his ministry with the Philadelphia Temperance Society, where Rush's teachings were considered gospel. While Graham was undoubtedly familiar with Rush's physiological warnings about overstimulation as the precursor to disease, something about the subject fired his imagination. His biographers explain that Graham, in his careful ledger notes, indicated that he purchased a fairly considerable medical library to continue his study of the theory of overstimulation of the nerves, inflammation of the digestive organs, and disease. His studies led him to the thesis that, in terms of the unhealthy effects of stimulating appetites on the body, the real culprit was gluttony. Stimulation, intemperance, overindulgence, and excessive appetites of all kinds were, as he would later advise in his massive book *The Science of Human Life* (1839), the greatest of all causes of evil.[14] To be constantly overstimulated was to open oneself to all kinds of salacious tastes and opportunities to sin, which in turn caused a cascade of health effects, including susceptibility to disease.

Graham believed, as many of his peers did then and as some do now, that the primary cause of overstimulation (and therefore disease) was sex. Specifically, the body was liable to fall prey to disease if an individual engaged in any sexual act that fell outside the Biblical consecration of sex for reproduction within marriage. Sex for pleasure (even within marriage) and masturbation were, to Graham, both the symptom and the contagion, and the subject of his famous sermon, "A Lecture to Young Men, On Chastity" (1838). Today, popular histories and magazine articles about the origin of the cracker that bears his name often focus on Graham's emphasis on sex as the logic behind his diet. It seems that the equation of sexual stimulation with overloaded nerves, which in turn cause disease, is too memorable to be passed over without comment.[15] But to only focus on the sensational element misses two important reasons for Graham's insistence on a diet of fruit, vegetables, homemade brown bread, and water. Graham had a genuine desire to aid disease prevention by teaching his followers to throw off the gluttony and unhealthy appetites that lead to physical illness. This abstention could lead to a number of positive outcomes, including everything from gaining a more equanimous temperament to the slackening of

[14]Graham, *Lectures on the Science of Human Life*, 19.

[15]One fairly recent example is Braun's (2014) *Atlantic* article "Looking to Quell Sexual Urges? Consider the Graham Cracker."

chronic complaints ranging from asthma to migraines to irritable bowel syndrome. In fact, at the beginning of his career as a health reformer, Graham's mission was concerned much less with sex and much more with holistic bodily and spiritual health.

Graham's forays into the spiritual and medical treatment of malaise, disease, and unhealthy predilections of the flesh and gut came at a fascinating time in American religious history. By the 1830s, the hardly half-a-century-old country was edging out of the Second Great Awakening, the religio-cultural movement that saw the spread of new Protestant Christian denominations throughout the nation. For nearly thirty years, a variety of Christian denominations—most famously, Methodism and the Baptists—swelled in numbers of adherents, and in many places surpassed older, more established churches in followers. The rigorous Calvinist theology that had dominated the pre-revolutionary period was slowly replaced by a revival of the workings of the Holy Spirit in the new nation and an "evangelical patriotism."[16] This flourishing of religious thought and practice was one of the American peculiarities observed by the French diplomat Alexis de Tocqueville, who remarked that religious fervor in America was common because it was "perpetually warmed by the fires of patriotism."[17] In the religious free market economy, purveyors of religion had to compete to recruit followers, and, once they had those followers, inculcated in their newly recruited souls a zeal for creating the Kingdom of Heaven in the USA. The glut of religious suppliers in the spiritual marketplace was possibly what kept Graham out of church and forced him to make a living first with a temperance society, and then as a lecturer: There were many thousands of ministers, but considerably fewer outlets for their ministry.

It was partially due to the inundation of religious thought that Americans turned to reform movements such as temperance leagues or health campaigns. Historian Ruth Engs observes as religious awakenings in America waned that a religious revival like the Second Great Awakening became "a generalized and powerfully felt cultural critique."[18] Engs describes several "cycles of health reform" in American history, and notes that as each religious reform movement has faded it is

[16] McLoughlin, *Revivals, Awakenings, and Reform*, 106.

[17] Quoted in Pessen, 75.

[18] Engs, 2001, 2.

often followed by a "clean living movement," of which Graham's crusade is prototypical. In some ways, it seems that after the market for *religious* reform has been saturated, Americans felt the need to turn instead to *cultural* reform, not unlike Graham's moral/dietary movement.[19] This cycle—a spiritual upheaval and renewed interest in religion, followed by a social reform movement—has happened several times in American history. These "awakenings" occurred in the eighteenth, nineteenth, and (according to some historians) twentieth centuries, and each time, Engs argues, a religious awakening is trailed by a clean-living movement. These clean-living movements, in turn, are accompanied by a related social/political program, during which legislative bans on substances like alcohol and tobacco are called for, or the physiological and moral necessity of abstinence or physical fitness become part of educational curriculum. Engs explains that, for the most part, health-reform movements "have emerged in the wake of, or tangential to, a revival of religious interest. This religious climate and cultural awakening have yielded an evangelical fervor among some individuals or groups who then strove to eliminate a perceived problem."[20] In other words, the soul might be unwilling to gird itself morally because the flesh is literally too weak, and thus a whole new reform movement is convened to solve what is now construed as a social problem.

Therefore, while the Second Great Awakening faded into a social reform movement in the early 1830s, Graham built his philosophy of health based on the writings and research of physicians, as well as his own spiritual tastes, so to speak. His work brought together contemporary scientific speculation with spiritual beliefs about the effects of food and alcohol on the human body. What Graham became best known for over the course of the 1830s was his assertion that health was the natural state of the body, and disease was due entirely to external forces. "Mankind ought to know that disease never results from the constitutional and legitimate operations of the human system," Graham wrote. "If, then, the body becomes diseased, it is always the result of some disturbing, some offending cause; and the disease can only be kept up while such a cause continues to prevent the healthy operations of the system; and health can only be recovered by the healthy operations of

[19] Engs 4.

[20] Engs 7.

the system."[21] Disease, in Graham's science, is provoked by unhealthy stimulation and inflammation of the "nerves of organic life." In order to resist disease, one must also resist that which interferes with the "condition of the stomach," because when "this organ is in good health, and properly supplied with healthful food, and the function of digestion is vigorously and healthfully going on, then it is that all the vital functions of organic life are most vigorously and perfectly performed...then has life the most complete victory over the causes which induce death."[22] Foods that interfere with the proper function of the stomach include the aforementioned alcohol, but also meat, coffee, tea, mustard, any spice except salt and pepper, and water that was too cold and from an unclean source. The stomach, Graham explained, "cannot suffer alone: being, as it were, a kind of retina or sensorium to the nerves of organic life, the whole system of those nerves, and consequently all the organ is supplied by them, sympathize powerfully in all its conditions and affections, partake of all its irritations, and suffer a consequent debility."[23]

The gut was, then, the location of all the health-giving and heath-blocking magic. The entire journey of food through the alimentary canal was a trip through what Graham several times refers to as "one sheet of sympathy." In short, in the Graham rubric of health, fruits, vegetables, homemade brown bread, and cool water not only prevented stimulation, but in so doing also curbed both disease and sinning.

The Asclepian Tablets of the Nineteenth Century

Graham's thesis was put to the test on the small scale of everyday sins—sexual as well as gluttonous—when in 1832 cholera, the scourge that had swept through the major cities of Europe, made landfall in New York City. Cholera was by that time a well-known and deeply feared public health emergency across India, the Middle East, and Europe. Scarcely a single major city had been spared the ruinous outbreaks that killed tens of thousands. And the terror engendered by the disease was well-earned. Within hours of the first symptoms, the infected are stricken by diarrhea and/or vomiting, both of which lead to intense stomach cramps.

[21] Graham, Introduction. *The Æsculapian Tablets*, VI.

[22] Graham, *A Lecture on Epidemic Diseases*, 12.

[23] Graham, Sylvester. *A Lecture on Epidemic Diseases Generally and Particularly the Spasmodic Cholera*, 13.

The purgative effects of the bacterium *Vibrio cholerae* results in catastrophic dehydration, which eventually leads to the circulation system shutting down. Even today, cholera is responsible for infecting 1.3–4 million and killing anywhere from 20,000 to over 143,000 people worldwide each year.[24]

The outbreak in North America began in Montreal at the beginning of June 1832, and by late June had traveled by sea, or possibly the recently completed Erie Canal, to New York City. Whether it was a single "patient X" or an entire ship's passengers disembarking from London or Paris, within weeks the disease had spread from the Old World throughout the New. Americans had been both mentally rationalizing and imaginatively dismissing the threat of cholera since it appeared in London in the previous January. As medical historian Charles Rosenberg remarks, the main outbreaks in Europe had been in the poorer sections of the large cities (the "slums") where "filth, misery, vice, and poverty conspired to produce [cholera's] unfortunate victims"; but in America:

> the healthy farmers and sturdy mechanics of the United States could, Americans believed, never provide such hecatombs of victims as cholera had claimed from among the pagans, Moslems, and papists of Europe and the East. America had no class to compare with the miserable slum-dwellers of Paris and London or with the brutalized serfs of Nicholas' Russia... 'With clean persons and clean consciences,' Americans were prepared to meet the disease without trembling.[25]

Added to this, doctors explained that cholera was "not a contagious disease, but may depend upon a change in the state of the atmosphere not cognizable by our senses, but which affects only those whose deranged digestive organs predispose them to the disease."[26] Cholera, the New York Board of Health explained, is transmitted by air, and in its aspirated form recombines with the neurobiological system of those who were already susceptible, where "susceptibility" was understood to mean weakness of some kind, including in morals. Thus, those with a predisposition to becoming ill were by and large assumed to have some sort

[24] Organization, World Health. "Cholera: Fact Sheet." *Media Center Fact Sheet*. WHO, http://www.who.int/mediacentre/factsheets/fs107/en/.

[25] Rosenberg, *The Cholera Years*, 15.

[26] *A Rational View of the Spasmodic Cholera*, 7.

of moral failing, and because poverty was a sure sign of lack of fitness (long before social Darwinism), if the poor residents of the slums sickened and died it was inevitably their own doing.[27] As Rosenberg puts it in describing the lack of preparedness by New York City's doctors: "Contagionism as a medical theory was frowned upon; invoking it could disrupt the social order, and it threatened the moral superiority of those who used the disease's prevalence among the poor as pretense for proselytizing."[28]

What happened when cholera reached New York City is an important moment in public health history. Over 3500 people died in the 1832 summer outbreak, and not only those who were poor: Many of the city's elites succumbed, as well. Fewer than half the people who were sickened recovered. The local government was overwhelmed by the ravages of the disease, as well as the ultimately predictable public protest that they had done nothing to make their citizens safe. Added to this was the fact that the evidence of John Snow's London water pump—which revealed that the cholera bacterium spreads via polluted water—would not be discovered for another twenty-two years. To add insult to the outbreak, the rich fled for the cleaner air of the country, and once again the burden of illness fell on the poor, crowded together in New York's tenements. Not coincidentally, in the aftermath, the social status of the victims of the disease were used as evidence of their susceptibility to illness and, consequently, their moral failings.

Into this maelstrom waded Sylvester Graham, for whom this outbreak proved to be pivotal to his career. Even before cholera reached America, Graham saw an opportunity to move from theory to practice, and to take the results of his physiological studies and put them to work in explaining how to fortify the body and the spirit against disease. Consolidating the theories of Rush and two French medical theorists—Xavier Bichat and François Broussais—Graham composed a public lecture that conflated the results of his own medical studies with his theological underpinnings. In March 1832, as word of the disease's effects in Europe was

[27] Our Physician explains: "It is generally admitted that filthiness, drunkenness, debauchery, bad food, want, excess, indigestion, scanty or improper clothing, cold, fear, fatigue and anxiety, are *predisposing causes*; and that although the atmosphere of the sick may sometimes form a focus of *infection*, yet even this is not dangerous except to such as have been subjected to one or more of those *causes*" (*A Rational View of the Spasmodic Cholera*, 8).

[28] Rosenberg, *The Cholera Years*, 77.

spreading throughout the cities of Northeast America, Graham gave his lecture on cholera at the Baptist meeting house in New York to a crowd of over 2000 people. The attendance was so extraordinary, and the sense of doom so great, that Graham was invited to repeat the talk several times throughout the summer up and down the East coast. In his "Lecture on Epidemic Disease Generally and Particularly the Spasmodic Cholera," Graham laid out his physiological theories—that "the inner surface of the stomach is more peculiarly the centre [*sic*] of sympathy to the whole organic system….so that every affection and every disturbance of the stomach, influences, in a greater or less degree, every organ and every function in the body"—and his broader religio-social beliefs in the efficacy of his diet of vegetables and brown bread for maintaining unstimulated virtue. Furthermore, Graham contended that panic at the prospect of an outbreak, or widespread belief that an infectious disease will become an epidemic, is sufficient to cause a community to "develop the same symptoms and cause the same results," and suggestions that his listeners remain calm, lest they induce the disease through the imaginative power of believing they are already infected.[29] With this as his foundation, Graham goes on to argue that we can build on this phenomenon of physical symptoms of a disease manifesting themselves even when there is no illness to understand how "certain actions of the mind"—a mind of unhealthy appetites or unfounded fears—"may produce certain conditions of the nervous system, causing corresponding effects on the organs and functions of the body…may so act on the bodily sympathies, and through them, on the organs and tissues, as absolutely to induce the same involuntary phenomena in most or all; and thus render them extensively epidemic."[30] A body made weak by continually succumbing to temptations such as tobacco, alcohol, rich foods, or the more salacious appetites, he advises, will have destroyed the organic nervous system, rendering the whole digestive system unable to extract from food the health-giving nutrients the body requires to function. As proof, Graham offers the examples of the waste laid by cholera in the heart of Paris, where the "drunken and the lewd have fallen almost by hundreds and thousands before this terrible destroyer."[31]

[29] Graham, *A Lecture on Epidemic Diseases*, 27.

[30] Graham, *A Lecture on Epidemic Diseases*, 29.

[31] Graham, *A Lecture on Epidemic Diseases*, 49.

"Let me not be misunderstood," Graham warned. "I do not say that mental action will absolutely induce spasmodic cholera, in any state of the body; but, that in certain conditions of the body, mental action may so affect the nerves of organic life, as to cause that gastro-intestinal irritation, which is the basis of all the symptoms of spasmodic cholera: and that continued mental action of the same character, may reduce the body, from any state of health, to that condition in which the disease may be superinduced."[32] In other words, to control the health of the mind and thereby maintain one's moral character would create an immunity to the first major epidemic in America's short history. "Every appetite and every passion," Graham explained—food, drink, and flesh—"should therefore be held in strict subjection to enlightened reason and moral propriety, if we would not increase our liability to be attacked by this terrible disease."[33] *Belief*—that sinful behavior abounds in the poverty-ridden slums, those "moral fens and morasses of society"—has mixed with *logic:* Residents of those tenements and slums, sinners who live in these morasses and refuse to change their evil ways will "become the sources of pestilence and death to those who have, perhaps culpably, neglected to improve their condition."[34] The only solution, then, is to purify the body, the mind, and the soul, with a diet free of stimulants and through the maintenance of "the utmost composure and serenity; and happy is that man who has that peace with God, which will enable him at all times, even in the hour of imminent peril, to cast himself upon the protection of his heavenly Father, with sustaining confidence."[35]

Other physicians preached precisely the opposite of Graham's advice, suggesting that a diet rich in animal flesh and alcohol would stave off the disease; others tended to agree with some of Graham's principles, but still believed that following his program to the letter would spell disaster. The effects of Graham's diet went unremarked on, however, until two years later when Graham published an extraordinary pamphlet made up almost entirely of grateful letters from men and women who had adopted his program. In "The Asclepian Tablets of the Nineteenth Century" (a reference to the Greek god of medicine, Asclepius), many "Grahamites"

[32] Graham, *A Lecture on Epidemic Diseases*, 32.

[33] Graham, *A Lecture on Epidemic Diseases*, 51.

[34] Graham, *A Lecture on Epidemic Diseases*, 36.

[35] Graham, *A Lecture on Epidemic Diseases*, 53.

extolled the diet not just for the physiological and spiritual benefits, but for facilitating their survival of the cholera outbreak. Some described how the diet made them virtually immune to the disease. David S. Burger wrote that while the disease was raging through the cities "I was very much among the sick of that terrible disease…I sat beside the sick, by the hour,—watched with them, rubbed them, lifted them, etc., yet through the whole cholera season, I had not the least touch of the complaint, nor the slightest indisposition of any kind."[36] Similarly, Anson Willis explained how he "spent the whole of the past summer in the city, and visited the sick of the cholera in the hospitals and elsewhere, and was much exposed; but had not the least symptom of that, nor of any other disease about me," while during the outbreak George Tracy "enjoyed the most perfect health; and while everybody else was full of dismay, we were perfectly composed, and without any apprehensions of danger in regard to ourselves."[37] Others, like James Little, exclaimed over the personal benefits of the diet, saying, "I had not followed your System long, before my health began to improve; and in a little while all my difficulties left me; and for more than seventeen months, I have not had an unwell hour. Last summer, during the whole cholera season, I walked every day three miles out, and labored all day at the carpenter's business, and then walked home at night; and through the season enjoyed excellent health."[38] Evander D. Fisher reported that, even though he lived at the outbreak's epicenter in New York City, he and his family "never enjoyed better health than we did through the whole cholera season." And each of the thirty-four letters echo some of the same sentiment expressed by John B. Jansen, who wrote that he is now "perfectly satisfied, that, as a general rule, every man may enjoy health if he will; and that it is disgraceful to be sick."[39]

As these contemporary "tablets to the great god Asclepius" show, Graham's followers believed that their improved constitutions and serene souls enabled them to avoid becoming sick. They believed that their soothed "sympathetic nervous system" inoculated them against the disease, making two misattributions of correlation as causation that

[36] Graham, *The Æsculapian Tablets*, 22.

[37] Graham, *The Æsculapian Tablets*, 54 and 80.

[38] Graham, *The Æsculapian Tablets*, 83.

[39] Graham, *The Æsculapian Tablets*, 92.

seem scientific but amount to superstition: (1) That their morally and physically superior bodies rendered them immune to the cholera, and not simply because they were doing things like drinking rainwater (as David Burger explains in his letter) or lived well away from outbreak sites (which seems true of James Little), and (2) That *all* followers of the Graham diet would meet the same fate. There is no mention of those who followed Graham's prescription but still fell ill (a situation which, after the fact, might be attributed to the cholera victim being morally compromised in some way). In this example of nineteenth-century dietetical practice, still loosely affiliated with scientific thinking, the inherent properties of things—specifically food and nutrients—are believed to have some form of moral ability to aid or to change or correct or corrupt. Fortunately, today we reap the health benefits of rigorous biomedical science and dietary choices that are the culmination of over a century of research on nutrition, human health, the gut's biome, and agricultural science. Thanks to the sciences of epidemiology and bacteriology, we now know that cholera is the result of consuming water polluted by *Vibrio cholerae*. The benefits brought by scientific research in the dietary fields cannot be overstated for the health it brings to human life.

Today, however, instead of the spiritual outcomes Graham was trying to inculcate on his herd, we are more focused on physiological outcomes as principles to live by. The social reform inherent in soothing the organic nerves that sustain the human gut has been replaced with the preventative treatments of diet and exercise: Science has supplanted spiritual and social cues. Yet, in many ways we can say that Graham's revolution lives on today, but this time in a scientific, rather than a religious, form.

Bibliography

Alcott, Louisa. "Transcendental Wild Oats." *The Independent … Devoted to the Consideration of Politics, Social and Economic Tendencies, History, Literature, and the Arts (1848–1921)* 25, no. 1307 (1873): 1569–71.

Austen, Jane and William Dean Howells. Pride and Prejudice. *The Modern Student's Library.* New York, Chicago: C. Scribner's Sons, 1918.

Braun, Adee. "Looking to Quell Sexual Urges? Consider the Graham Cracker." *The Atlantic*, 2014.

Brodsky, Alyn. *Benjamin Rush: Patriot and Physician.* 1st ed. New York: Truman Talley Books, 2004.

Francis, Richard. *Fruitlands: The Alcott Family and Their Search for Utopia*. Yale: Yale University Press, 2010.

Graham, Sylvester. *A Lecture on Epidemic Diseases Generally and Particularly the Spasmodic Cholera. With an Appendix Containing Several Testimonials,—Rules of the Graham Boarding House, & C. By Sylvester Graham*. Day, 1833.

———. *The Æsculapian Tablets of the Nineteenth Century*. Printed by Weeden and Cory, 1834.

McLoughlin, William Gerald. *Revivals, Awakenings, and Reform: An Essay on Religion and Social Change in America, 1607–1977. Chicago History of American Religion*. Chicago: University of Chicago Press, 1978.

Nissenbaum, Stephen. *Sex, Diet, and Debility in Jacksonian America: Sylvester Graham and Health Reform*. Contributions in Medical History Series No. 4. Westport, CN: Greenwood Press, 1980.

Organization, World Health. "Cholera: Fact Sheet." *Media Center Fact Sheet*. WHO. http://www.who.int/mediacentre/factsheets/fs107/en/. Accessed 29 June 2018.

Pessen, Edward. *Jacksonian America; Society, Personality, and Politics*. The Dorsey Series in American History. Homewood, IL: Dorsey Press, 1969.

Physician. *A Rational View of the Spasmodic Cholera: Chiefly with Regard to the Best Means Preventing It*. Boston: Clapp & Hall, 1832.

Rosenberg, Charles E. *The Cholera Years the United States in 1832, 1849, and 1866*. Chicago: University of Chicago Press, 1987.

Rush, Benjamin. *Medical Inquiries and Observations Upon the Diseases of the Mind*. The History of Medicine Series No. 15. Published Under the Auspices of the Library of the New York Academy of Medicine by Hafner Pub. C., 1962.

Salinger, Sharon V. *Taverns and Drinking in Early America*. Baltimore: Johns Hopkins University Press, 2002.

Starr, Paul. *The Social Transformation of American Medicine*. New York: Basic Books, 1982.

Index

M. Mathias and A. M. Moore (eds.), *Gut Feeling and Digestive Health in Nineteenth-Century Literature, History and Culture*, Palgrave Studies in Literature, Science and Medicine,
https://doi.org/10.1007/978-3-030-01857-3

E

CPSIA information can be obtained
at www.ICGtesting.com
Printed in the USA
LVHW080143091218
599509LV00011BA/223/P

9 783030 018566